Management of Civic Energy and the Green Transformation

An increase in the global demand for energy, combined with an increase in the price of energy and energy products, has advanced the growing interest in renewable energy technologies and the wide implementation of renewable energy sources (RES). Member States of the European Union have been global leaders in the use of renewable energy and in the transition to new technologies. *Management of Civic Energy and the Green Transformation: A Case Study of Poland* examines the current issues of transitioning from traditional energy sources to newer, renewable energy sources, while balancing supplies, and working synergistically with existing, conventional sources.

Features:

- Offers a balanced blend of theory and practice of development economics for renewable energy implementation
- Presents a case study of how Poland is working towards their energy transition, and provides other examples and recent statistical data from other European Union countries
- Analyses the legal and systemic conditions supporting the development of renewable energy systems and offers direction on the potential for the green development of the civic energy sector

Management of Civic Energy and the Green Transformation

A Case Study of Poland

Anna Brzozowska, Piotr Maśloch,
and Grzegorz Maśloch

CRC Press is an imprint of the
Taylor & Francis Group, an **informa** business

First edition published 2023
by CRC Press
6000 Broken Sound Parkway NW, Suite 300, Boca Raton, FL 33487–2742

and by CRC Press
4 Park Square, Milton Park, Abingdon, Oxon, OX14 4RN

CRC Press is an imprint of Taylor & Francis Group, LLC

© 2023 Anna Brzozowska, Piotr Maśloch, and Grzegorz Maśloch

Reasonable efforts have been made to publish reliable data and information, but the author and publisher cannot assume responsibility for the validity of all materials or the consequences of their use. The authors and publishers have attempted to trace the copyright holders of all material reproduced in this publication and apologize to copyright holders if permission to publish in this form has not been obtained. If any copyright material has not been acknowledged please write and let us know so we may rectify in any future reprint.

Except as permitted under U.S. Copyright Law, no part of this book may be reprinted, reproduced, transmitted, or utilized in any form by any electronic, mechanical, or other means, now known or hereafter invented, including photocopying, microfilming, and recording, or in any information storage or retrieval system, without written permission from the publishers.

For permission to photocopy or use material electronically from this work, access *www.copyright.com* or contact the Copyright Clearance Center, Inc. (CCC), 222 Rosewood Drive, Danvers, MA 01923, 978–750–8400. For works that are not available on CCC please contact *mpkbookspermissions@tandf.co.uk*

Trademark notice: Product or corporate names may be trademarks or registered trademarks and are used only for identification and explanation without intent to infringe.

ISBN: 978-1-032-44082-8 (hbk)
ISBN: 978-1-032-44083-5 (pbk)
ISBN: 978-1-003-37035-2 (ebk)

DOI: 10.1201/9781003370352

Typeset in Times
by Apex CoVantage, LLC

Contents

List of Figures ix
List of Tables xi
Contributors xv

Introduction 1

              Overview of the Book 3

Chapter 1   Energy in the Face of Socio-economic and Environmental Changes .. 5

          1.1   Problems of the Development of Modern Energy 5
          1.2   The Role of Energy in Socio-economic Development
                 and Its Impact on the Natural Environment............................ 12
          1.3   Creative and Innovative Approaches to Energy in Modern
                 Societies and Economies 16
                 1.3.1   Clean Coal Technologies............................... 17
                 1.3.2   Smart Grids 17
                 1.3.3   Energy Storage 21
                 1.3.4   New Technologies for Renewable Energy Sources... 22
                 1.3.5   Technologies Seeking to Eliminate Low Emissions . 22
                 1.3.6   Prosumer Energy 23
                 1.3.7   E-Mobility 23
                 1.3.8   New Business Models (Organisational, Legal
                         and Financial) 24

Chapter 2   Efficiency in Energy 27

          2.1   The Concept and Essence of Efficiency 27
                 2.1.1   Traditional Approach..................................... 28
                 2.1.2   Resource-based Approach.............................. 28
                 2.1.3   Evaluation of Adopted Strategies.................. 28
           2.2   Measurement of Energy Efficiency 30
           2.3   Energetics Efficiency and Energy Efficiency......................... 35

Chapter 3   Costs in the Energy Sector 43

          3.1   Characteristics of Costs in the Energy Industry and Their
                 Classification 43
           3.2   Factors Determining Costs in the Energy Industry................. 47
                 3.2.1   Energy Externalities 47
                 3.2.2   External Costs Versus Externalities.............. 50
           3.3   Internalisation of External Costs in the Energy Industry 52
                 3.3.1   Pigou Tax... 53

		3.3.2	The Coase Theorem 54
		3.3.3	Methodology for Determination of External Costs... 55
	3.4	Instruments for Internalising Costs 59
		3.4.1	Environmental Taxes and Charges 59
		3.4.2	Voluntary Agreements 62
		3.4.3	Ecological Compensation... 62
		3.4.4	Fiscal Reform 62
		3.4.5	Deposit Fees 64
		3.4.6	Financial Penalties 64
		3.4.7	Direct Regulation 64
		3.4.8	Instruments Based on Market Transactions 65
		3.4.9	Subsidies 67

Chapter 4   Managing Transformation in the Area of Civic Energy 69

4.1   Energy Transition Historically and Now 69
4.2   Effects and Challenges of the Current Energy
         Transformation—Strategic Directions of Development.......... 74
4.3   Change Management and the Role of Leaders 81

Chapter 5   Conditions for the Development of Renewable and Civic Energy..... 85

5.1   Definition, Division and Characteristics of Renewable
         Energy.. 85
         5.1.1   Re a) Solar Energy 86
         5.1.2   Ad b) Wind Energy 87
         5.1.3   Ad c) Geothermal Energy 87
         5.1.4   Re d) Hydropower Energy 87
         5.1.5   Ad e) Biomass.. 88
5.2   The Importance of Renewable Energy for the Economy 88
5.3   The Impact of Renewable Energy on the Natural
         Environment and Human Health 93
5.4   Influence of Social Awareness on the Possibilities
         and Directions of Development and Use of Renewable
         Energy Sources 95
5.5   Directions and Trends in the Development of Renewable
         Energy—the Key Role of Civic Energy 97

Chapter 6   Renewable Energy in the Polish Energy Sector—Resources
                      and Use 105

6.1   The Energy Sector in Poland... 105
6.2   Resources, Potential and Use of Renewable Energy 111
         6.2.1   Solar Energy.. 113
         6.2.2   Hydropower... 114
         6.2.3   Wind Energy.. 115

Contents

vii

		6.2.4	Geothermal Energy	115
		6.2.5	Biomass	116
	6.3		Legal Basis for the Use of Renewable Energy Sources in Poland in Terms of the Development of Civic Energy	119

Chapter 7 Analysis of the Development of Civic Energy in Poland 123

	7.1	Introductory Remarks	123
	7.2	Description of the Study	125
		7.2.1 Interviews with Representatives of Municipal and District Authorities	130
		7.2.2 Interviews with DSO Management	132
		7.2.3 Interviews with Residents/Entrepreneurs	133
	7.3	Stakeholders Involved in the Implementation of Projects and the Potential Costs and Benefits of Participating in the System	133
		7.3.1 Society	134
		7.3.2 Enterprises	134
		7.3.3 Local Authorities	135
		7.3.4 State as a Regulator	137
	7.4	Organisation of an Energy System Development Model Based on Local and Regional RES	138

Conclusion 145

Bibliography 149

Index 163

Figures

1.1	Environmental Kuznets curve	9
1.2	The interplay between technology, industry and market in the renewable energy sector	15
2.1	The procedure for transforming the various efficiency categories into economic efficiency	31
2.2	Energy inputs, energy services and GDP	32
3.1	The impact of negative externalities on the welfare changes of the three primary market participants: producers, consumers and the state	49
3.2	Pigou tax	53
3.3	LCOE of RES variables and fossil-fuel power plant operating in load base from 2017 to 2050 (in Euro/MWh for r = 7%)	58
3.4	LCOE results for EU27 in 2018 (in Euro 2018/MWh)	58
4.1	Growth in primary energy demand (TWh) from 1850 to 2019	71
4.2	Share of energy from renewable sources in EU countries (as % of gross final energy consumption)	72
4.3	Degree of electrification of the mature EU economy by 2050	72
5.1	Share of Modern Renewable Energy in 2020 (in %)	98
5.2	Global Investment in Renewable Power and Fuels, 2011–2021	99
7.1	Responses to question 1: What forms of civic energy do you know? (More than one answer could be selected)	125
7.2	Responses to question 4: From which sources was the RES investment financed (more than 20% of investments—more than one source could be indicated)	127
7.3	Responses to question 6: Who do you think should be the leader of change in the implementation of RES technologies in your municipality/region	129
7.4	Responses to question 7: What are the main barriers to the development of civic energy in Poland?	129
7.5	Responses to question 8: What forms of public sector support are key to the development of civic energy? (More than one answer could be indicated)	130

7.6 Responses to question 10: Do you have a group of trusted people (e.g. 10) among your immediate neighbours/co-workers with whom you would be able to make an effort to build organised civic energy structures (e.g. an energy cooperative) 131

7.7 Analysis of connections in the local civic energy structure presented using the ARA model as an example 140

Tables

1.1	Fossil fuel potential at the end of 2020	6
1.2	World total energy consumption by region and fuel (in BTU)	8
1.3	Basic directions of influencing the improvement of the natural environment in terms of production and consumption	11
1.4	Incentive and pricing schemes used in DSR	19
1.5	Classification of DSR programmes	21
2.1	Efficiency dimensions and their characteristics	30
2.2	Breakdown of forecasting models in the energy economy	35
2.3	Age structure of the housing stock in Poland built before 2002 and its unit energy demand indicators	37
2.4	Comparison of EPC and EDC contracts	41
3.1	Summary of energy technology features	45
3.2	Unit investment costs for energy generation technologies (EUR/kWe)	46
3.3	Highly efficient organisation model	47
3.4	Criteria for qualifying a liability as an environmental tax	60
3.5	Environmental tax categories	61
4.1	First and second energy transitions	70
4.2	SWOT analysis in relation to the prosumer segment	78
4.3	SWOT analysis carried out for the community segment	79
4.4	SWOT analysis for the local government segment	80
5.1	Definitions of renewable energy	86
5.2	Socio-economic effects of renewable energy use	89
5.3	Main problem areas related to renewable energy	90
5.4	Global renewable energy employment by technology in 2020 (thousands)	92
5.5	Jobs directly created in renewable energy during construction and installation, as well as exploitation and maintenance phases	93
5.6	Energy payback time (EPT) by technology	95
5.7	Global potential of renewable energy resources (in EJ)	97

5.8	Potential instruments that can be used by public authorities for energy policy planning	101
6.1	Energy consumption (in Mtoe)	106
6.2	Production and consumption of primary energy in the years 2012–2020	106
6.3	Selected data on the operation of NPS from 2015 to 2021	107
6.4	Household electricity consumers and consumption	108
6.5	District heating infrastructure and sales of thermal energy	108
6.6	ODEX indicator, base year 2000 = 100	110
6.7	Energy efficiency indicators in households sector in 2010, 2015 and 2020	110
6.8	Impact of selected factors on variation of final energy consumption in the years 2010–2020 (Mtoe)	111
6.9	The technical potential for energy obtainable from renewable energy sources in Poland per year	112
6.10	Share of energy from renewable sources in total final energy consumption (%)	112
6.11	The share of renewable energy sources in the total renewable energy obtained in the years 2016–2020	113
6.12	Average annual values of sunshine in hours for selected cities of Poland	114
6.13	Solar energy—photovoltaics	114
6.14	Theoretical and technical potential of Polish rivers	115
6.15	Hydropower (in TJ)	115
6.16	Capacities of power stations using wind	116
6.17	Geothermal energy	117
6.18	Capacities of power stations using biomass (MW)	118
6.19	Gross final energy consumption from renewable sources in the years 2016–2020 (in TJ and in %)	119
7.1	Responses to question 2: Do you have RES, if yes which ones? (more than one source could be selected)	126
7.2	Responses to question 3: What was (will be) your primary and secondary purpose for the RES investment?	126

7.3	Responses to question 4: From which sources was the RES investment financed (more than 20% of the investment—more than one source could be indicated)	127
7.4	Responses to question 5: What form of citizen energy do you represent?	128
7.5	Responses to question 9: Are you interested in further developing civic energy—if so, in what form?	130

Contributors

Anna Brzozowska, PhD, is an associate professor in the field of economics in the discipline of management science. Currently, she works in the Department of Logistics at the Faculty of Management of the Częstochowa University of Technology. Since 2000, her research and teaching activities have been associated with the Częstochowa University of Technology. She is the author or co-author of about 180 scientific publications. Her research interests include the issues of management, transport, logistics processes and innovative pro-ecological solutions, the subject of which she addresses in her scientific publications. From 2018 to 2021, she was the editor-in-chief of Scientific Journals of the Częstochowa University of Technology Management.

Piotr Maśloch, PhD, is an associate professor at War Studies University in Warsaw, Management and Command Faculty. He is an officer of the Polish Army and Dean's Representative for EU projects. He is also an author and manager of international projects (founded by UE) in the field of new technologies and modern education. He is an active teacher and researcher in management, sustainable energy, sustainability and environmental impact. He is also a reviewer of scientific publications of the best Polish universities and international publishing houses, as well as an author or a co-author of more than 100 publications as paper contributions in international journals, book chapters and conference proceedings. His research interests include management, new technology, global aspects of management and globalisation.

Dr. Grzegorz Maśloch is Assistant Professor in the Department of Economics and Finance at the Warsaw School of Economics. He is the head of postgraduate studies for Local Government Managers and Head of Postgraduate Studies of Waste Management at the Warsaw School of Economics. His research interests include the issues of strategic management and investment projects implemented in renewable energy, as well as the issues of innovative pro-ecological solutions and energy efficiency.

Introduction

The world economy's growing need for energy, together with rising energy and energy carrier prices, has resulted in an increasing interest in renewable energy and the use of renewable energy sources (RES). Renewable energy sources are nowadays an alternative to traditional, primary, non-renewable energy carriers (fossil fuels). RES are replenished by natural processes, which make them practically inexhaustible. An unquestionable advantage of obtaining energy from these sources is the low level of negative impact on the environment. For the development of efficient, sustainable and effective energy based on renewable resources, civic energy is becoming crucial.

Civic energy is defined as a system in which individuals, organisations, institutions and companies outside the energy sector play and active role in energy generation, transmission and management. An issue characteristic for civic energy is its local, regional and small-scale production of electricity and heat from renewable sources and the use of energy efficiency solutions. The hallmarks of civic energy, on the other hand, are voluntary and open membership, participation in the profits of investments and activities of entities operating in the local energy market, which are owned, managed and controlled democratically by local communities.

The energy sector in Poland is currently facing new major challenges. The demand for energy, growing from year to year, with an outdated production and transmission infrastructure, dependence on external supplies of energy carriers in the form of natural gas and oil, as well as the deteriorating state of the natural environment and international obligations in the area of environmental protection make it necessary to take decisive, systemic actions leading to key changes in the energy sector. This applies to many spheres of economic and social activity, including, in particular, organisational, investment, educational and scientific activities (Polityka energetyczna, 2009).

As the energy stability of a country determines the possibilities and directions of its development, this problem is particularly important in Poland, a country with a highly energy-intensive economy and low level of energy efficiency. Faced with the need for the energy transition, it is necessary to look for new solutions not only in the technological sphere but also, or perhaps mainly, in the social, political and organisational spheres. In this respect, civic energy may be helpful, which, by including society in the transformation processes, allows for the integration and involvement.

It should also be kept in mind that civic energy production tends to involve small-scale generators located close to the consumer, thus increasing local energy security and reducing transmission losses. Renewable energy production is characterised by low emissions, ensuring positive environmental effects. RES can also make a significant contribution to the national energy balance, as well as to the energy economy of municipalities, districts or even regions. RES should also contribute to energy security and especially to improving energy supply in areas with underdeveloped energy infrastructure.

DOI: 10.1201/9781003370352-1

The possibility of obtaining energy from RES leads to a qualitatively new situation, both in the spheres of energy production and distribution and on the side of end consumers. From the point of view of the hitherto traditional way of energy production and distribution within the established systems, the emergence of RES energy makes it necessary to take additional energy into account in balancing production and supply. As a result, RES leads and will lead to a reduction in demand for non-renewable energy, and therefore a reduction in consumption of fossil energy. It can therefore be assumed that such a mechanism will force a reduction in the negative environmental impact and change the rules of functioning for all forms of conventional energy.

It is also obvious, and understandable, that the aforementioned mechanism of substitution of dirty energy by clean energy will lead to negative technical and economic effects in the companies involved in the production and distribution of conventional energy. It is to be expected that there will be a deterioration of the economic account of these enterprises, influenced by an increase in production costs, especially fixed costs, with a concomitant reduction in sales revenues due to the reduction in production. This effect may be mitigated by an increase in energy demand, which will, to some extent, flatten out the impact of the negative effects. However, this argument is too weak to pose a threat to the development of RES systems. These systems have already proven their suitability and effectiveness in many countries for energy production and supply. The only target solution, therefore, can be the construction of energy systems based on the coexistence of conventional energy production and distribution and RES systems, with an indication of a gradual, systematic increase in the production and consumption of renewable energy.

The development of civic energy based on renewable energy also entails a number of phenomena that were previously unknown in energy systems. The technical, technological or organisational solutions associated with RES offer completely new possibilities that are not yet known in the professional energy sector. In particular, these are the possibilities of building local and regional energy subsystems that can achieve high degrees of energy independence. The construction of subsystems using dispersed energy producers that are also a source of power creates real opportunities for consumers to become at least partially independent of energy from producers and distributors operating within conventional systems. The development of such subsystems is already fully feasible today. This is determined by the available technology and techniques for generating and distributing energy in a manner that is completely independent of the aforementioned conventional energy. Without going into organisational and technical details, it should be assumed that energy generation from RES can be diversified to produce energy in quantities that fully meet local and regional needs.

In the course of the conducted research and literature reviews, a number of technical, organisational, financial, economic and social problems were recognised which, to varying degrees, affect the course of RES implementation processes in Poland. The scope of problems in this study therefore includes new and hitherto unrecognised research areas, which significantly condition the socio-economic and environmental effects of RES implementation.

This study addresses technical, organisational, social, legal, political, environmental and, to a greater extent, financial and economic issues. These are the problems

Introduction 3

that are given the most attention in this study, because in the authors' opinion, disregarding the environmental or technical conditions, it is the economic dimension, and thus the social dimension, that is the sine qua non condition for the effective diversification of energy production and supply in Poland.

The main objective of this study is to analyse the possibility of implementing systemic solutions for civic energy in Poland, which would allow energy needs to be met at a socially useful and acceptable level.

OVERVIEW OF THE BOOK

The first chapter is devoted to energy and energy-related issues and their importance in the modern world. Issues concerning the impact of energy and the energy sector on social and economic development are considered. The environmental impact of energy and new approaches to energy issues in highly developed countries are also discussed in detail.

Chapter 2 addresses the issue of how efficiency can be defined, interpreted and measured in the energy sector. In this part of the study, a definitional distinction is made between energetics efficiency and energy efficiency.

Chapter 3 discusses issues concerning the characteristics of costs and the factors that determine them. The chapter presents conceptual issues and classification of costs in the energy sector. The chapter also presents the factors determining costs in the energy sector and opportunities for cost internalisation.

Chapter 4 deals with the problems of managing the energy transition—with a particular focus on the civic energy aspects. Some information on the energy transition is presented, including a historical perspective. In addition, challenges of the ongoing second energy transition are discussed.

Chapter 5 considers the socio-economic determinants of renewable energy development. In this section, renewable energy is characterised, along with its features and division. An important part of the chapter is a consideration of the impact of renewable energy on the environment and human health, as well as on the economy and society. Development of the energy industry takes place in a specific social space, which is crucial for the technical and organisational solutions adopted in the energy industry. Public awareness therefore determines the possibilities, directions of development and use of RES, which is also another issue addressed in the chapter.

Chapter 6 is devoted to the role of renewable energy in the Polish energy sector. The resources, potential and use of RES are characterised in detail, divided into the main groups: solar energy, hydropower, wind power, geothermal power and biomass.

Chapter 7 presents an analysis of the potential and opportunities for civic energy development. The issues were considered taking into account the local and regional resources, as well as individual stakeholders and their potential participation in civic energy development projects. This part of the study discusses possible activities and initiatives, implementation of which may contribute to the sustainable development of civic energy in Poland.

The comprehensive analysis of problems concerning renewable energy in the aspect of civic energy development, which is discussed in this study, is particularly significant nowadays, in the face of the energy crisis. The concepts presented in the study, expanded by the authors' own analyses and conclusions, may constitute a valuable input in a discussion on the future of energy and energy security in the world. As the authors demonstrate, the second energy transition is primarily about abandoning fossil fuels as soon as possible in favour of renewable energy sources and a green deal. Success of the green energy transition, however, will depend on the extent to which civic energy is developed and integrated into the transition processes. Only such approach, emphasised and promoted by the authors, must be implemented in the near future in the consciousness of individual societies, on the assumption that there is simply no other way to further sustainable development of the global world.

1 Energy in the Face of Socio-economic and Environmental Changes

CONTENTS

1.1 Problems of the Development of Modern Energy 5
1.2 The Role of Energy in Socio-economic Development and Its Impact on the Natural Environment...12
1.3 Creative and Innovative Approaches to Energy in Modern Societies and Economies..16
    1.3.1 Clean Coal Technologies 17
    1.3.2 Smart Grids...17
    1.3.3 Energy Storage..21
    1.3.4 New Technologies for Renewable Energy Sources 22
    1.3.5 Technologies Seeking to Eliminate Low Emissions.........................22
    1.3.6 Prosumer Energy...23
    1.3.7 E-Mobility...23
    1.3.8 New Business Models (Organisational, Legal and Financial)...........24

1.1 PROBLEMS OF THE DEVELOPMENT OF MODERN ENERGY

The energy sector produces, using various resources and methods, the unique good (commodity) of energy. Energy, which still cannot be stored in industrial quantities, is consumed on an ongoing basis and demand and supply are strongly dependent on the weather. The aforementioned characteristics of energy indicate that its development and use depend on a number of socio-economic and environmental conditions.

Today, economies around the world are, and will continue to be for decades to come, significantly dependent on fossil fuel-based energy. As estimates show (see Table 1.1), the potential of fossil fuels is still significant. The world's largest fossil fuel reserve is coal, which at current consumption levels, will be sufficient for about 100 years. In contrast, much smaller reserves exist for oil and gas, which are sufficient for about 50 years (Banks, 2015, pp. 401–409; Chow et al., 2003, p. 1528). In terms of current consumption, there are still huge reserves of fossil fuels that guarantee the possibility of global economic development in the coming decades. It should also be kept in mind that new discoveries and ongoing technical advances in the extraction or processing of fossil fuels may yet significantly improve these possibilities (Chow et al., 2003, p. 1528).

DOI: 10.1201/9781003370352-2

TABLE 1.1
Fossil Fuel Potential at the End of 2020

Oil—Total proved reserves

	Thousand million tonnes	Share of total	R/P ratio
OECD	38.3	15.0	25.2
Non-OECD	206.1	85.0	66.9
OPEC	171.8	70.1	108.3
Non-OPEC	72.6	29.9	24.5
European Union	0.3	0.1	10.1
US	8.2	4.0	11.4
Total World	**244.4**	**100.0**	**53.5**

Coal—Total proved reserves at end 2020

	Total—Million tonnes	Share of total	R/P ratio
OECD	508,433	47.3	363
Non-OECD	565,675	52.7	90
European Union	78,590	7.3	266
US	248,941	23.2	*
*More than 500 years			
Total World	**1,074,108**	**100.0**	**139**

Natural gas—Total proved reserves

	Trillion cubic metres	Share of total	R/P ratio
OECD	20.3	10.8	13.7
Non-OECD	167.8	89.2	70.6
European Union	0.4	0.2	9.2
US	12.6	6.7	13.8
Total World	**188.0**	**100.0**	**48.8**

Source: Own elaboration adapted from BP Statistical Review of World Energy, BP, 2021, pp. 6, 34, 46.

Note: R/P—an indicator of the ratio between the resources of a given raw material R (*reserves*) and the production level of a given fuel type P (*production*). The R/P ratio indicates the probable exploitation period of these resources in years, at current production levels.

Most of the recognised and available fossil fuels remain the responsibility of underdeveloped countries. For developed countries (and especially the European Union), the situation is much more complicated. Most of these countries have relatively few resources at their disposal, which makes them directly dependent on fossil fuel imports.

As the possible exploitation period of a raw material reserve is reduced, the price of fuel derived from it tends to increase. Therefore, the demand for fuel derived from a given raw material decreases and is replaced by other, alternative types of fuel. This process tends to create conditions in which a given fossil fuel can be used for a longer period than current research indicates. Fossil fuels will therefore continue to be exploited intensively over the next decades. As numerous studies have shown, the world's non-renewable energy resources, according to the Hubbert curve, are around or have even exceeded the global peak curve, indicating that half of these resources, or even more, have been consumed. It is therefore indisputable that fossil fuel reserves are finite, and that current energy consumption and growth patterns are not sustainable in the long term (Twidell and Weir, 2015, p. 3).

Furthermore, the use of energy from conventional sources is incompatible with sustainable development goals, among others, resulting in:

- combustion of fossil fuels, which has a serious negative impact on the environment (especially air in the atmosphere),
- harmful emissions from conventional means of power generation that have a negative impact on human health,
- limited development opportunities for a significant part of the world's population that does not have access to energy services,
- increasing inequalities in access to energy that contribute to social instability,
- energy security issues for countries/regions that make significant use of imported energy,
- loss of opportunities to make the full use of raw materials in the future (contrary to the principles of a circular economy).

There are also many problems with nuclear power. Although the merits of nuclear power are generally not questioned, extensive research in many countries show that the public approach the operating of nuclear power plants in their regions with reluctance and apprehension, and a large majority is strongly opposed to new projects. The main concerns relate to the risk of accidents and the threat of terrorism (Global Public Opinion, 2005, pp. 10–23). The economic costs of redeveloping sites after closing nuclear power plants are also significant (Fujita, 2015, pp. 7–12). This leads to a situation where nuclear power is considered unsafe in many countries around the world and decisions are made to close more sites. Confirmation of this assumption can be found in the nuclear power development scenarios, conceptualised by organisations supporting its development (e.g. International Atomic Energy Agency), which, even under the most favourable circumstances, do not foresee an increase in its rank in the global energy balance (Moriarty, 2012, p. 245). On the other hand, however, it should be kept in mind that in countries that do not have sufficient resources of their own energy, there are often calls for the development of nuclear power. This is the case, for example, in many EU countries, where the case for nuclear energy has been strengthened in particular by Russia's aggression against Ukraine. Advocates of nuclear energy, present it, among other things, as a way to make Europe independent of Russian gas and oil imports while reducing emissions.

Energy demand is an important authentic indicator depicting the development of humanity in a given region, in terms of its fundamental aspects (longevity, skills, innovation, standard of living, etc.) (Rana and AlHumaidan, 2016, p. 32). Socioeconomic development is therefore closely linked to an increase in energy consumption, which is fostered by, among other things, a growing world population, rising living standards and ongoing technological advances. This results (Krawiec, 2010, pp. 15–16) in an increased demand for scarce energy resources and natural resources, a transfer of resources to highly developed countries and widening of the development gap between rich and poor countries, and a progressive loss of biodiversity.

TABLE 1.2

World Total Energy Consumption by Region and Fuel (in BTU)

Region and fuel	2020	2030	2040	2050
Total OECD				
Liquid fuels	82.3	89.0	88.9	91.9
Natural gas	68.0	71.2	73.8	78.4
Coal	26.7	23.1	23.1	22.3
Nuclear	19.3	19.2	17.2	13.8
Other	36.7	48.8	60.8	74.9
Total	**232.9**	**251.3**	**263.8**	**281.3**
Total	**368.6**	**453.9**	**531.7**	**605.1**
World				
Liquid fuels	182.4	215.6	231.6	248.5
Natural gas	147.3	166.5	178.1	193.2
Coal	155.6	155.8	168.3	176.7
Nuclear	27.5	31.7	33.5	32.7
Other	88.7	135.6	183.9	235.2
Total	**601.5**	**705.2**	**795.4**	**886.3**

Source: Own elaboration adapted from Reference case, Annual Energy Outlook, 2021.

An analysis of global energy consumption projections clearly indicates that energy demand is expected to continue to grow rapidly in the coming decades (Goldemberg, 2006, p. 6; Luukkanen et al., 2019, pp. 8–39; Energy to 2050, 2003, pp. 19–56). As can be seen from the projections presented in Table 1.2, global energy consumption will increase steadily in the coming decades. World energy consumption by energy source in 2020 and projections to 2050 are shown in Table 1.2.

An analysis of the world's energy consumption in historical and forecast terms shows a huge disparity between OECD and non-OECD countries. If, in 1990 or 2000, it was the OECD countries that consumed significantly more energy than the other countries, by 2012 the situation had already changed dramatically. Moreover, forecasts clearly indicate that energy consumption in non-OECD countries will grow much faster than that of OECD countries. Energy consumption between 2020 and 2050 is projected to increase globally from 601.5 BTUs to 886.3 BTUs (by 47.4 per cent), of which among OECD countries from 232.9 BTUs to 281.3 BTUs (by 20.8 per cent). Non-OECD countries will increase their energy consumption by as much as 64.2% (from 368.6 BTUs in 2020 to 605.1 BTUs in 2050).

Projections of global consumption by individual source clearly indicate a growing demand for energy, which will be produced mainly from fossil fuels in the perspective of the year 2050. In the case of nuclear energy, production will be maintained at current levels. Globally, renewable energy will develop most dynamically and play an increasingly important role in securing the world's energy needs.

As energy demand increases as a result of population growth and economic growth (Chien-Chiang and Chun-Ping, 2007, pp. 1206–1223; Chiou-Wei et al., 2008, pp. 3063–3076), savings in energy consumption can only be achieved by increasing energy efficiency. However, a problem arises here, defined in the literature as the 'rebound effect' or 'Jevons paradox'. According to this concept, the introduction of new technologies or organisational solutions to increase energy efficiency brings with it certain side effects expressed in terms of increased consumption. In practice, this means that when energy efficiency is achieved in some area of economic or social activity, the savings are usually reinvested in economic or social activities elsewhere, which results in a net effect, that is, an increased energy consumption (Pieńkowski, 2012, pp. 105–116).

The fundamental problem is therefore not only the technological use of new energy sources but also the change in consumption patterns to ones that do not increase energy demand (Pienkowski, 2012, p. 106).

So energy efficiency needs to be looked at in a much broader perspective, where the 'rebound effect' is just one element to be considered. More importantly, it is also the element that in most cases leads to increased prosperity. This is particularly relevant in relation to one of the outcomes of the 'rebound effect', which is an increase in innovation and productivity. In some cases, it may also be the case that externalities resulting from the 'rebound effect' are purely beneficial (Gillingham et al., 2014, p. 24).

Consideration of energy cannot be undertaken without taking into account its impact on the environment. The theoretical approach depicting the impact of human activity caused, inter alia, by the use of energy resources on the environment is reflected in the environmental Kuznets curve (Jankowska, 2016, p. 54) (see Figure 1.1), according to which there is a relationship between well-being and devastation and use of natural resources.

According to the theory proposed by Kuznets, there is a saturation point in every economy and society, beyond which increasingly more attention is paid to the state and quality of the environment. It is therefore possible to draw the conclusion that highly developed countries that achieve a high level of economic development and guarantee a high quality of life for their inhabitants pay increasing attention to issues

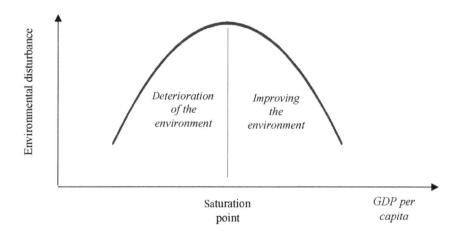

FIGURE 1.1 Environmental Kuznets curve.

Source: Own elaboration adapted from T. Żylicz (2015, p. 125).

related to environmental technologies and behaviour. This is determined both by economic policy and by the pressure of a society that increasingly appreciates the issues related to a high quality of life, for which access to safe and clean energy and a well-maintained and unpolluted environment is essential.

In the early stages of economic development, environmental pollution and consumption of natural resources increase, but after a certain level of per capita income (the saturation point, which can vary depending on many factors) is reached (Brzozowska et al., 2021c; Stern, 2004, pp. 1419–1439), it starts to decrease. Thus, at high per capita incomes, economic growth leads to environmental improvements and reduced consumption of natural resources (Stern, 2004, p. 1419). The environmental Kuznets curve thus indicates that underdeveloped countries are more likely to accept greater environmental pressures compared to highly developed countries, which are more likely to undertake pollution reduction initiatives (Jankowska, 2016, p. 55).

The first stage of the environmental Kuznets curve is confirmed by empirical studies, while the subsequent stage of the curve is based solely on hypotheses and its probability is linked to the degree of compliance with recommendations derived from the sustainable development strategy. Therefore, the approach reflected in the environmental Kuznets curve is widely criticised (Stern et al., 1996, pp. 1151–1155).

An analysis of the results extensively reported in the literature on the impact of energy on the development of the environmental Kuznets curve (de Bruyn, 1997, pp. 485–495; Bruvoll and Medin, 2003, pp. 27–48; Suri and Chapman, 1998, pp. 195–208) indicates that a key factor influencing the possibility of reducing emissions generated by the energy sector is technological progress. Technological changes are, of course, occurring in all countries; nevertheless, the key ones are emerging and are first introduced in highly developed countries and only later adapted by less developed countries (Stern, 2004, pp. 1419–1439). In addition, countries remaining at a lower level of development bear the environmental costs of relocating companies and locating industries, often with less innovation and high energy requirements (consequently generating additional pollution and waste) (Suri and Chapman, 1998, pp. 205–206).

There is a risk of international tensions due to differing needs, approaches to global problems in energy production and consumption and obvious conflicts of interest. As an analysis of historical statistics on energy use and energy efficiency shows, most of the world's highly developed countries have, with their socio-economic development, increased and continue to increase their energy demand. Consequently, as the quality of life in highly developed countries increases, so does the environmental cost of energy consumption. This has led to a situation in which a serious ethical issue arises as to whether highly developed countries have the right to impose strict norms and standards on the extraction and use of energy, and whether, as a result, they have the right to prohibit developing countries from extracting the necessary amount of energy with the same extraction techniques that they themselves used until recently.

However, underdeveloped countries face dilemmas as to whether to follow the path taken by highly developed countries, with all the ecological consequences this entails, or, on the contrary, to take a development leap and immediately apply clean, environmentally friendly technologies (Burchard-Dziubińska, 2007, p. 231; Klein, 2016, p. 425). This is, of course, debatable, and the energy revolution that would take place bypassing the development of conventional energy requires breaking down the

Energy in the Face of Socio-economic and Environmental Changes   11

knowledge barrier. However, as an analysis of the telecommunication transformation in Africa, where the landline development stage was virtually skipped in many regions, such revolutions are possible and effective.

Highly developed countries are implementing policies aimed at reducing the use of fossil fuels through a variety of measures. They are therefore taking measures to increase the share of energy production from renewable sources or to increase energy efficiency and energy-saving technologies. The implementation of stricter norms and rules regarding energy production and consumption and its impact on the environment very often leads to 'carbon leakage' (Jankowska, 2016, p. 61). The consequence of such practices can be the relocation of the most emission- and energy-intensive production to underdeveloped countries with lower standards in terms of working conditions or the environment.

The fact that market prices in most countries (even highly developed ones) do not, in any way, consider the externalities and adverse effects that are a consequence of the use of fossil fuels remains unsolved (Song, 2006, p. 6). This therefore leads to an obvious situation where the market price of energy derived from fossil fuels is significantly lower than energy generated from available alternative energy sources. At the same time, it should be made clear that the implementation of measures to improve environmental standards in the energy sector will entail enormous economic costs. A choice will also have to be made in the social sphere between material wealth and environmental wealth. There are five basic possible courses of action for environmental improvement in terms of energy production and consumption (Bekkeheien et al., 1999, pp. 95–96). In Table 1.3, the basic directions of influencing the improvement of the natural environment in terms of production and consumption are presented.

TABLE 1.3
Basic Directions of Influencing the Improvement of the Natural Environment in Terms of Production and Consumption

Directions	Impact
Conscious sacrifice	It consists of voluntarily giving up consumption or consciously giving up certain solutions and habits (e.g. giving up over-consumption of goods or services, reduced demand for housing/residential needs or heating and cooling).
Clean fuels	Development of technologies that allow the comprehensive use of cleaner fuels (e.g. hydrogen) in the economy.
Reduction of emissions	Taking action to clean or level the emissions generated. Emission reductions can use primary methods (increasing the efficiency of equipment, treating the fuel, removing pollutants from the fuel before combustion, using clean combustion technologies) or secondary methods (exhaust gas cleaning or levelling by capturing).
Improving energy efficiency	Improving energy efficiency can be implemented in all areas of human functioning. It concerns the sphere of production, service provision, transport, public administration or the functioning of households.
Change of fuel	Switching to environmentally friendly energy sources (e.g. RES).

Source: Own elaboration.

The implementation of the aforementioned lines of influence on energy production and consumption can take place in one way or, as is most often the case, is a mix of the elements mentioned. The implementation of each option entails specific, very often significant costs. However, these costs must be incurred in order to implement an energy mix in economies that will enable societies to develop in a safe and sustainable manner. The degree of technological advancement and available technologies is undoubtedly a key in this respect. However, one should not forget the political or social factors that determine the degree of acceptance of innovative and environmentally friendly solutions.

1.2 THE ROLE OF ENERGY IN SOCIO-ECONOMIC DEVELOPMENT AND ITS IMPACT ON THE NATURAL ENVIRONMENT

Development can be considered in economic, social and environmental terms. As societies develop their economies, they seek to improve their quality of life, which requires securing certain energy needs. Energy is produced and used in a specific environment. Nowadays, the challenge is to build a relationship between society, economy and the environment in which sustainable development is possible.

Energetics and the energy it produces are essential to the functioning of societies and economies, and the ways in which it is used affect all aspects of socio-economic life, as well as political or environmental issues. Energy also underpins the economic and social development of modern states. It is the driving force behind technological and economic development and has a significant impact on the standard and quality of life. However, through the use of fossil fuels, the consumption of which is and will remain the main source of energy, representing one of the world's greatest environmental impact dilemmas. As energy is essential for the internal and external security of a region/state, energy-related issues are fundamental challenges with a direct impact on the functioning and development of societies and economies.

Economic development and an increase in the well-being of society are therefore made possible by an efficiently and effectively functioning energy sector. The energy sector is widely recognised as a 'key sector' with a significant impact on both the development and competitiveness of other economy areas on society's quality of life. The energy sector is widely considered to be highly centralised and characterised by features inherent in a natural monopoly, and the services and products it provides are characterised as public goods. It also has a key influence on the degree and type of infrastructure provision in a given region or country.

There is a widespread view in the literature that it is impossible to implement the principle of sustainable development without introducing sustainable energy management (Pieńkowski, 2012, p. 110). Undoubtedly, economic development and environmental problems cannot be considered separately, but must be analysed as processes with a high degree of interdependence. This means, therefore, that the implementation of sustainable development goals will be possible in a green economy, a success of which will be determined by sustainable energy.

Sustainable energy is a way of producing and managing energy that ensures sustainable access to sufficient energy for present and future generations, while limiting the negative effects of its production and management on the environment (including climate change). Notwithstanding some discrepancies in defining sustainable development and the role energy plays in it (Robert et al., 2005, pp. 8–21), virtually all current energy systems in the world fail to meet the defining objectives. This is determined by the fact that power generation is responsible for more than 80% of greenhouse gas emissions emitted by human activities. Energy consumption is steadily increasing, and global energy production is more than 80% dependent on fossil fuels (Goldemberg, 2010, p. 100). It is clear that the choice of energy sources depends on geography, economic situation, resource availability and security of supply. Environmental concerns about gas emissions and global warming are forcing a gradual shift away from the use of non-renewable energy resources. Therefore, the prices of energy resources depend not only on their availability or energy value but also on their environmental impact (Rana and Al Humaidan, 2016, p. 31).

The basic principle of sustainable energy is the efficient use of available energy, human, economic and environmental resources. The implementation of this principle takes place in a socio-economic and political environment in which certain patterns of consumption, behaviour and preferences have been formed. The effects of the implementation of sustainable energy principles include environmental issues (e.g. reduction of pollution, improvement of ambient air quality) and the organisation and functioning of society and the economy (e.g. access to and use of energy, cultural and political conditions, energy and public security).

An important change in the implementation of sustainable energy principles is brought about by the increasing attention (especially in highly developed countries) to quality-of-life issues. Quality of life is understood as the degree to which all the needs of society (both tangible and intangible) are met. The expectations raised by societies concern an ever-widening range of services or adherence to values. Therefore, public authorities, enterprises or individual citizens are required to provide increasingly higher quality of products offered and standards of services provided, and to respect values, including, among others, pro-ecological and pro-environmental attitudes.

An important consequence of implementing the principles of sustainable energy is the recognition that continued economic growth and progressive globalisation are not the best ways to achieve higher living standards and social well-being if it consequently leads to inequality, hidden costs and greater vulnerability of local economies to global restructuring and externally induced economic shocks. The new economy therefore favours the development of decentralised social and economic organisations and local government, proposing an evolution from the modern international economy, to an elite, decentralised and multi-level global economic system (Seyfang, 2009, p. 49).

An energy system can be considered sustainable when energy is supplied in increasing volumes to meet increasing socio-economic needs, while maintaining a relationship whereby the substances released as a result of energy production and consumption do not exceed the natural capacity of the environment to assimilate them.

The various, very often, different approaches to energy problems in individual countries or regions of the world result in situations where often opposing interests clash. On the one hand, it is widely acknowledged that fossil energy will continue to dominate for several decades to come. This is determined by the relative prices of fossil fuels, which, by not taking into account all external costs or opportunities foregone, are and will remain relatively inexpensive for years to come. In addition, it is important to be aware that conventional energy is supported by an extensive infrastructure of mines, pipelines, power stations, railways, petrol stations, tankers, vehicles, etc. The status quo is also guaranteed by the oil and energy corporations, one of whose objectives is to inhibit initiatives (technical, technological, financial or organisational) that could threaten their market position. In addition, any changes in conventional energy have a significant impact on public sentiment, especially in regions where mining functions or conventional energy production and services are predominant. Consequently, there are very strong political and economic centres around the world that ensure the protection and continuity of conventional energy investments (Chow et al., 2003, p. 1531). On the other hand, however, the possibility of maintaining energy consumption and securing energy demand growth with non-renewable resources is not sustainable in the long term. Renewable energy sources will become widespread when they become more competitive in relation to fossil fuel energy. By more competitive, we should primarily mean the relative price ratio. It is therefore clear that, instead of waiting idly and reflexively for fossil energy prices to rise, countries or regions can support the development of renewable energy. Measures to support RES development can be implemented through two coordinated, complementary and mutually reinforcing actions (Twidell and Weir, 2015, p. 3):

- first, by implementing policies and initiatives to promote RES, research and development activities. It is also important to undertake RES investment initiatives within the competence of the public sector. In the long term, pro-environmental and pro-innovation education is also extremely important, at all levels, from preschool and elementary school to promoting and highlighting good practices and green consumption patterns among seniors. These activities are mainly implemented by subsidising investments that will reduce the price and improve the cost-effectiveness of renewable energy compared to fossil fuels.
- second, the public sector can influence the price level of fossil fuels through fiscal instruments (e.g. carbon taxes) or environmental levies. This type of action, on one hand, increases attractiveness of investments in renewable energy sources and, on the other hand, signals the environmental costs caused by the consumption of fossil fuels and introduces them into the economic accounts. These measures thus consist of supporting investment in renewable energy sources by subsidising their development, while at the same time reducing the attractiveness and price competitiveness of fossil fuel energy.

J. Rifkin similarly focuses on the transformations taking place in the energy industry, writing about the third industrial revolution and the coming post-carbon era and its five pillars, which included (Rifkin, 2011, pp. 9–106):

- development of renewable energy,
- equipping buildings with micro-installations to consume the renewable energy generated in the building,

Energy in the Face of Socio-economic and Environmental Changes

- the use of energy storage technology in households,
- the use of network management and sale of surplus energy,
- transport based on innovative fuels (e.g. helium fuel and electricity).

These factors indicate that, under current conditions, the current models of energy sector operation need to start systematically changing and increasing their efficiency. Although conventional power generation will continue to dominate for many years to come, it is already necessary to start building innovative models of power generation, where renewable energy sources will increasingly take part in the energy mix (Brzozowska et al., 2021a). This must take into account the relationships that exist between technology, industry and market. The dynamic development of technology in the renewable energy sector is translating into more and more advanced technologies commonly available on the market, and thus into more and more cost-effective installations. New technical, organisational or legal solutions necessitate the continuous improvement of technologies and the undertaking of further research and development. The process of implementing new technologies also affects new markets and sectors, for which renewable energy is becoming increasingly accessible and attractive. Figure 1.2 presents the interplay between technology, industry and market in the renewable energy sector.

Renewable energy creates a self-perpetuating developmental system. As presented in Figure 1.2, the research and development work leads to the creation of new technologies, which are followed by their implementation in economic practice. The widespread use of innovative solutions in the field of renewable energy contributes to the emergence of new research problems both in the technological sphere and in the organisational, legal, financial or social spheres. Solving emerging problems implies new research and development initiatives, which are a source

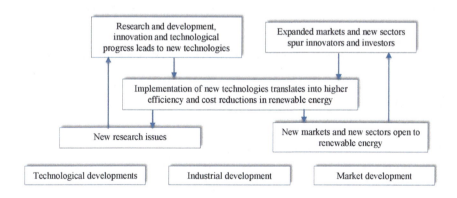

FIGURE 1.2 The interplay between technology, industry and market in the renewable energy sector.

Source: Own elaboration adapted from Renewable energy sources and climate change mitigation. Summary for policymakers and technical summary. Special report of the intergovernmental panel on climate change. Intergovernmental Panel on Climate Change, Potsdam, 2012, p. 151.

of further innovation and technological progress. However, new technologies affect new markets and industries that see renewable energy as a development opportunity. New and expanded markets, together with new industries involved in renewable energy initiatives, stimulate innovators and investors who engage in the process of implementing new technologies.

The construction of a modern model for renewable energy development and its cooperation within national energy systems requires the consideration of all parties and factors that have or may have any influence on its construction, including especially those parties that may operate within a civic energy framework. There are many interactions and reactions between parties and factors that influence the effective development of renewable energy (Song, 2006, p. 6). The number of factors and the possibility of interaction determine that the functioning and actions taken in renewable energy cannot be implemented without considering all aspects and effects that renewable energy can generate in the economy or society. Taking into account as many variables is essential for the development of renewable energy strategies and models.

1.3 CREATIVE AND INNOVATIVE APPROACHES TO ENERGY IN MODERN SOCIETIES AND ECONOMIES

As already established, the state of socio-economic development of any region or country depends on the degree of efficiency of the energy industry. Therefore, energy is an area of particular interest to those responsible for its organisation, production and functioning, as well as consumers of energy. It is also becoming an area where key innovations are emerging. The dynamic development of technology, new forms of organisation, financing or influencing social attitudes contributes to the creation of new ideas and products, which are introduced not only in the energy sector but also in production and energy-related services.

Energy is one of the most innovative areas of the economy. Energy innovation, by the very definition and characteristics of innovation, has evolved over time. For example, it is worth mentioning J.A. Schumpeter (Schumpeter, 1960, p. 104) and his broad definition of innovation[1] and the different view of the problem by P.F. Drucker,[2] J. Friedmann,[3] R.W. Griffin[4] or, for example, P. Kotler.[5] Similarly, energy innovation can be considered from different perspectives and scopes. Regardless of the scope of analysis, however, it is indisputable that the creativity and innovativeness of societies lead to constant changes in the energy mixes of individual countries or regions. Key aspects of research in modern economies concerning the energy sector are activities related to:

a) clean coal technologies,
b) smart grids,
c) energy storage,
d) new renewable energy technologies,
e) technologies seeking to eliminate 'low emissions',
f) prosumer energy,
g) E-mobility,
h) new business models (organisational, legal and financial).

Energy in the Face of Socio-economic and Environmental Changes

1.3.1 CLEAN COAL TECHNOLOGIES

Clean coal technologies are all processes and uses of coal aimed at minimising the negative environmental impact of its combustion products. The term therefore refers to the entire 'coal chain', from extraction to the disposal of the residue after its use.

The development of clean coal technologies is directly linked to the need to achieve better energy efficiency of coal, as well as to achieve economic efficiency that is competitive with other energy sources. However, as economic experience shows, the implementation of clean coal technologies poses a number of problems, and their application on a larger scale is unlikely to be possible in the coming years (Mianowski, 2012).

1.3.2 SMART GRIDS

In the case of electricity, it is crucial to guarantee a supply with highly stabilised parameters. Producers of energy supplied to the grid must adhere to very strict rules regarding the quantity and quality of the electricity supplied. On the one hand, as a result of an increase in demand for electricity, the dispatcher should be prepared to ensure that more electricity is supplied to the grid, while preventing possible voltage in case of decreased supply of electricity.

Only the latest electronic and radio technology has allowed for the development of the *smart grid* concept, enabling a significant reduction in power that the system should have at its disposal in order to meet the needs of consumers. A smart grid can be defined as an electricity network that is capable of harmoniously integrating the behaviour and actions of all users connected to it—generators, consumers and those who perform both roles—in order to provide a sustainable, economical and reliable power supply (Skoczkowski and Bielecki, 2015, p. 88). The pace and extent of smart grid development depends on (Gungor et al., 2013, pp. 28–42; Skoczkowski and Bielecki, 2015, p. 88):

- new technologies for the generation of electricity and heat, where renewable energy sources play a key role,
- the functioning dispersed sources and energy storage capacities,
- information technology to develop new methods and systems for controlling, monitoring and securing networks,
- the availability and popularity of devices for smart grid integration,
- new energy market development models to enable the development of dispersed energy, including prosumer energy,
- changes in energy consumption habits, aiming to use energy efficiently.

The *smart grid* concept involves integrating renewable energy sources to feed into the public power grid. The integration of new installations that load the power grids requires the management of variables as well as individual energy sources. Smart grids must therefore be characterised by completely different features than those inherent to traditional solutions. These features undoubtedly include effective energy management, rapid response to changing customer needs, equal treatment of all energy market participants, the ability to seamlessly connect additional users, and security of supply (Pamuła and Papińska-Kacperek, 2012, p. 63). Such perspective therefore forces market participants to adopt a completely new approach and behaviour. While in the currently commonly functioning system, the role of the energy consumer is minor and limited to signing a contract and paying for energy consumption, in case of smart grids, the role of the consumer is substantially different. The essence of a smart grid system is the active participation of consumers, who can both consume and produce energy. Smart grids should therefore enable, through smart management, the integration of intermittent sources (from renewable energy sources), including dispersed energy generating and storage technologies (Bielecki and Skoczkowski, 2014, p. 172).

The conscious and responsible behaviour of energy consumers has a significant impact on the management of the entire electricity grid and, in business practice, can already be successfully used to improve its efficiency, increase quality, reduce costs and integrate dispersed energy sources (Grycan et al., 2014, pp. 230–231). The active participation of end-users adopted in the *smart grid* system functions under the concept of integrated energy demand (DSI) and requires them to change their existing behaviour by, among other things, actively managing their energy consumption and supplying energy from renewable sources or storing it (Matusiak et al., 2011, p. 89).

Effective energy demand-side response (DSR) is also important and should soon start to play an increasingly important role in electricity system balancing. DSR is defined in various ways in the literature. The International Energy Agency (IEA) defines DSR as the ability of the demand side to respond to changes in electricity prices in the energy market or in real time. The main objectives of DSR include (Rasolomampionona et al., 2010, pp. 138–143):

- reduction of maximum peak loads,
- increased load during valley periods when energy prices are low,
- shifting loads between different times of day or seasons,
- matching the load to the current operating conditions of the electricity system.

The effectiveness of DSR measures can be considered in the short term or long term. In the short term (several hours), the demand side response affects the power balance in the system and represents the economic optimisation of electricity demand. In the long term, in addition to its impact on the energy balance, it can also be analysed through energy savings.

DSR programmes can be divided into mandatory and voluntary, and into incentive (incentive-based-IBP) and pricing (tariff-based-PBP) programmes (Albadi and El-Saadany, 2008, pp. 1989–1996; Rasolomampionona et al., 2010, pp. 138–143). A comparison of examples of incentive and pricing programmes used in DSR is presented in Table 1.4.

TABLE 1.4
Incentive and Pricing Schemes Used in DSR

Incentive programmes (IBP)

Programme	Characteristics
Direct load control (DLC)	A programme involving the manual or remote shutdown of the consumer's equipment over a specified cycle. The decision to switch off consumption results from conditions that threaten the reliability of the electricity system. The programme belongs to the group of mandatory programmes, adherence to which forces the consumer to reduce energy consumption each time (in exchange for a specific gratification) as required by the network operator.
Interruptible/curtailable rates (ICR)	A programme consisting of a contract with the customer, where a clause is established whereby the customer accepts interruption of all or part of the power taken or agrees to reduce the power taken at the supplier's request. The ICR programme is usually dedicated to the largest industrial customers.
Demand bidding programmes (DBPs)	A programme that creates load reduction incentives for large energy consumers at a price for which they are willing to reduce. By participating in the programme, electricity consumers commit to provide a load reduction service when certain operating conditions of the electricity system occur. The programme usually establishes the number of load reduction calls, the period during the day and the total number of hours of load reduction over a specified period.
Emergency demand response programmes (EDRPs)	The programme is voluntary. As part of it, the energy consumer has the option of deciding to participate in the energy curtailment on a given day. The programme involves the reduction of load by the electricity consumer on the recommendation of the transmission system operators. Consumers inform the operators of the amount and price at which they are prepared to reduce their consumption by submitting an offer, which is subject to evaluation and, consequently, once accepted, can be accepted. If, following an agreement, the consumer fails to reduce its load, it may be penalised.
Capacity market programmes (CMPs)	Programmes under which, in the event of certain operating conditions on the power system, customers commit to reduce load by a predefined amount. These types of programmes operate through periodic auctions where participants submit price bids for load reduction. The programme usually establishes the number of reduction calls, the duration of the reduction and the total number of hours of reduction within a defined period.
Ancillary services market programmes (ASMPs)	The programmes allow customers to submit offers to reduce load, in the regulation market, thereby increasing the scope of the available operating reserve. If customers' offers are accepted, they are priced according to the market price and paid for the standby commitment. If load reduction has to be used, the operator calls for the reduction and the service can be paid for according to immediate market prices.

(Continued)

20   Management of Civic Energy and the Green Transformation

TABLE 1.4
Incentive and Pricing Schemes Used in DSR (Continued)

Pricing (tariff) programmes (PBP)

Programme	Characteristics
Time-of-use (TOU) tariffs	A tariff in which the energy charge varies on a daily, weekly and seasonal basis. There are significant price differences, allowing the demand for electricity to be shifted from a period of high price for that energy (e.g. daytime peak) to a period of low price (e.g. night-time valley), thus limiting the negative effects of excessive demand. The impact of the tariff on consumers is stronger when the spread between rates for different times of the day, week or season is wide and when there is the stronger the wider the spread between rates for different times of the day, week or season and when there is the possibility to programme electrical equipment to operate in the load valley. A key element of the tariff is also the need to provide the electricity consumer with access to information on current consumption and price.
Critical-peak pricing (CPP) tariffs	In order to link TOU tariffs more effectively to the current operating conditions of the electricity system, one or two additional very high rates shall be introduced for peak load periods when prices on the wholesale electricity market are highest. Customers shall be informed in advance that these rates will be applied. The number of days when the CPP replaces the standard rate is limited.
Real-time pricing (RTP) tariffs	A tariff in which electricity prices are forecast to fluctuate over time. The electricity tariff fluctuates in a similar way to wholesale market prices, with customers being informed of forecast energy prices 1 hour to 24 hours in advance. The tariff adds transmission and distribution costs and the supplier's margin to the cost of energy.

Source: Own elaboration adapted from Rasolomampionona et al. (2010, pp. 138–143), Boisvert et al. (2007, pp. 53–74), Motowidlak (2017, pp. 1159–1161) and Cortés-Arcos et al. (2017, pp. 19–31).

The incentive and pricing programmes available under DSR are applied considering the response time required for effects to appear from their application. In addition to the time aspect, criteria related to the controllability of individual DSR programmes are also important, as well as the type of incentive or end-users they can target (see Table 1.5). As can be seen from the analysis of the programmes, they are mainly targeted at large industrial or commercial customers. The potential for DSR programmes to be used by individual consumers or SMEs is still much more modest.

Summarising the discussion on DSR, it is clear that a more comprehensive use of the various programmes can bring many benefits. Positive effects can be seen in case of energy consumers (e.g. increased energy efficiency, greater awareness of consumer energy consumption) and most importantly, in case of the electricity system itself and consequently, the country. In the case of the electricity system, the key benefits are increased security and stability of the electricity system, and in the case of the country, increased energy efficiency of the entire economy and reduced levels of CO_2 and other pollutants.

Energy in the Face of Socio-economic and Environmental Changes

TABLE 1.5
Classification of DSR Programmes

DSR programme	Classification criterion										
	Type of incentive		Traffic management / Operative			Control type			End-user		
	Incentive	Price	Reliability	Economical	Non-operational	Active	Passive	Tariff	Large industrial customer	Commercial customer	Private individuals and SMEs
Direct load control—DLC	X		X			X				X	X
Interruptible/ curtailable rates—ICR	X		X				X		X	X	
Demand Bidding Programmes—DBP	X			X			X		X	X	
Emergency demand response programmes—EDRP	X		X				X		X	X	
Capacity market programmes—CMP	X		X				X		X	X	
Ancillary services market programmes—ASMP	X		X				X		X	X	
Time-of-use tariffs-TOU		X			X			X	X	X	X
Critical-peak pricing tariffs—CPP		X			X			X	X	X	X
Real-time pricing tariffs—RTP		X			X			X	X	X	X

Source: Own elaboration adapted from Motowidlak (2017, p. 1162).

1.3.3 ENERGY STORAGE

The concept of energy storage has a long tradition and was developed in parallel with the emerging energy systems. The earliest solutions mainly used water, which, when stored in reservoirs or pumped to different levels, made it possible to produce electricity at the desired time. Nowadays, various energy storage technologies exist and are still being improved, which we can categorise according to their capacity. Large storage facilities are mainly used in conventional power generation, where they are used, among other things, to stabilise the frequency in transmission networks. Small storages are used in energy systems where it becomes necessary to balance energy production from uncontrollable renewable sources. Micro-storage or customised storage is also starting to play an increasingly important role, allowing prosumers to balance energy production at a time that suits them (Alotto et al., 2014, pp. 325–335; Diouf and Pode, 2015, pp. 375–380).

In the next decades, energy storage will become increasingly popular, with renewable energy operating in a distributed system becoming the main supplier of energy storage. As a result, the dynamic development of technology will lead to an increase in the capacity and efficiency of energy storage while reducing the size of energy storage tanks (BP Statistical Review of World Energy, BP 2021, p. 17). Moreover, in the future electricity grid, energy storage systems may be considered as major sources of electricity. Energy storage as a disruptive innovation in the power sector will change the future rules of the power system. (Tan et al., 2021, pp. 1–3). The development of storage technology and its widespread application and use will also create entirely new opportunities for prosumer and civic energy.

1.3.4 New Technologies for Renewable Energy Sources

Renewable energy is one of the fastest growing economic sectors in the world. It is estimated that the share of renewable energy in energy balances, in terms of electricity, heat and fuels consumed in transport, will continue to grow rapidly in the coming years (Global Trends in Renewable, 2017). The trends taking place in renewable energy are the result of a paradigm shift in the global energy industry—marking a gradual shift from conventional energy towards dispersed energy, in which renewable energy plays a significant role. New technologies that are increasingly efficient and accessible play an important role in this transition.

1.3.5 Technologies Seeking to Eliminate Low Emissions

Low emissions are understood as emissions of pollutants emitted from sources at low altitudes, that is, primarily pollutants associated with human activity, most often emitted by individual household cookers and stoves, small boiler houses and transport. Low-lying emission sources cause high concentrations of pollutants that have a negative impact on the quality of life and health of people, the technical condition of buildings and infrastructure, and all forms of animate and inanimate nature (Maśloch, 2015, pp. 102–105).

A low-carbon economy is also an economy that can grow without increasing greenhouse gas emissions. The low-carbon economy is primarily based on increased energy efficiency, the use of renewable energy sources and the widespread adoption of innovative technologies, mainly in industry, construction, transport and agriculture.

Effective reduction of low emissions is possible through measures aimed at, among other things:

- thermo-modernisation of buildings that remain in a poor state of repair,
- replacing low-efficiency heat sources (mainly low-quality fuel) with environmentally friendly boilers or renewable energy sources,
- reducing car transport and promoting public transport,
- the gradual replacement of conventional energy by renewable energy sources.

Energy in the Face of Socio-economic and Environmental Changes 23

1.3.6 PROSUMER ENERGY

Prosumption is developing in many sectors of the economy, mainly in innovative fields using the latest technologies. One such sector is the energy sector, which is currently undergoing a number of developmental changes involving, inter alia, a paradigm shift from the traditional model based on large energy producers towards dispersed energy. This is mainly due to the dynamic development of renewable energy sources or technology that enables intelligent energy management and control. The aim of prosumer energy is to create a situation in which the energy consumer is both producer and consumer at the same time. Any surplus energy generated by the production of energy for their own needs (electricity or heat) can be sold by the prosumer to other consumers. The significance of prosumer energy in this case as a disruptive innovation lies in the fact that the prosumer makes 'investment' decisions in an environment that integrates (via value chains) supply and demand (Maśloch, 2016, pp. 198–199).

When analysing the functioning of prosumers in the energy sector, it should be kept in mind that, in addition to the conditions for individual prosumers, particular interest should be given to prosumer groups. A community of prosumers (prosumers grouped according to, e.g., similar energy sources and similar amount of energy supplied to the grid) is, in contrast to individual prosumers, a community which, by bringing together parties with similar interests or behaviour, is able to achieve common goals on a much larger scale and scope (Rathnayaka et al., 2011, p. 199).

The development of prosumer energy is also a reference to the 'big push' concept of P.N. Rosenstein-Rodan, according to which it is necessary for economic development to achieve a certain amount of basic capital, the characteristics of which are high investment and low variable costs (Jakubczyk, 2010, pp. 134–135). Prosumers, by building networks and using savings or profit generated, can proceed to co-create core capital, which in the future will contribute to the creation of new enterprises.

1.3.7 E-MOBILITY

A comparison of the efficiency of petrol (internal combustion) and electric engines clearly shows the advantage of the electric car. Of course, what is a prerequisite for e-mobility is the requirement that the electricity used to power electric vehicles must not come from coal-fired power stations. In a situation where energy derived from coal combustion is used to power electric cars, what can be gained through the use of an electric motor will be lost along the efficiency chain comprising: boiler, turbine, generator and individual voltage networks.

The development of the electric vehicle market worldwide depends on a number of factors, including (Electric Vehicles in Europe, EEA, 2016, p. 51):

- scope of CO_2 regulation,
- financial incentives for the purchase and exploitation of electric vehicles,
- favourable treatment of electric vehicles in given areas
- fuel prices,
- battery costs,

- accessible public transport,
- the wealth of societies and the investment capacity of businesses,
- infrastructure to enable seamless use of electric vehicles,
- trends and travellers' preferences.

1. The number of electric vehicles (EVs) sold has been increasing exponentially in past years. In 2021, sales of electric cars (including all-electric and plug-in hybrids) doubled compared to 2020 and reached an all-time record of 6.6 million cars. The total number of electric cars sold at the end of last year was around 16.5 million cars. According to the Global Electric Vehicle Outlook report, another record is likely to be beaten in 2022 (Global EV Outlook, 2022. Securing supplies for an electric future, IEA Publications, 2022, pp. 4–7).

1.3.8 New Business Models (Organisational, Legal and Financial)

There is no doubt at all that energy must now be used rationally. However, the level of its consumption is treated as a measure of civilisational and economic development. Therefore, it becomes important to look for new organisational, legal and financial models to ensure the development of new creative solutions in economic practice. This is by all means a complex issue, requiring not only the resolution of technical, legal or financial issues, but—usually more difficult—social or political problems. Such a holistic view of the determinants of energy development necessitates the search for new, innovative solutions in the organisational and legal spheres, as well as proposing financial models adapted to particular types of investment.

The reasons for the increasing interest in the search for new organisational, legal and financial models are primarily to be found in Brzóska and Krannich (2016, p. 8) and Burger et al. (2014, pp. 301–391):

- a transparent concept for creating value for the customer, the prosumer and the business owners,
- looking for instruments and methods to achieve a variety of competitive advantages (cost advantages, quality features),
- creating a structure that is capable of generating value as a basis for generating income and contributing to energy security,
- the use of the business model as a vision of the business idea, offered to potential investors and lenders,
- treating the business model as a carrier for different types of innovation,
- increasing public awareness, including in particular the importance of and opportunities for civic energy development,
- the development of civil society and the improvement of activities in terms of launching new initiatives and building social networks.

Extremely important from the point of view of energy development is the course of the innovation diffusion process, which follows a logistic curve. In the initial period, the process of adoption of innovations is relatively slow, in order to start a

Energy in the Face of Socio-economic and Environmental Changes

rapid process of dissemination of solutions after some time, which consequently leads to a high level of saturation, after which the focus shifts to other solutions and the process of adoption of new solutions can start again at another time. The process of innovation diffusion triggers far-reaching changes of a socio-economic nature. It means the dissemination of achievements resulting from the invention and creativity of a relatively small group of creators (innovators) to a wide audience. The recipients of innovations transferred through distribution channels here gain specific external benefits. This is because many of them do not participate in the creative process leading to innovative solutions, but precisely through the diffusion mechanism they become participants and consumers in this process (Jarosiński, 2015, pp. 31–33). The mechanism of diffusion of innovations is clearly visible precisely in the energy sector, where new solutions are created by a few, and later adopted or used by many entities (enterprises, public administration institutions, households, etc.).

The implementation of sustainable energy principles depends, to a significant extent, on the degree of involvement of individual citizens in the investments, that is, mainly on the degree of civic energy development.

NOTES

1 According to J.A. Schumpeter, innovation arises in the following cases: the manufacture of a new product or the introduction of goods with new characteristics to the market, the introduction of a new production method, the opening of a new market, the acquisition of new sources of raw materials and the carrying out of a new organisation of economic processes (see Schumpeter, 1960, p. 124).
2 P.F. Drucker regarded every novelty as an innovation. In his view, innovation did not have to be something technical or material (see Drucker, 1993, p. 18).
3 J. Friedmann believed that innovation is the result of a creative act, resulting in the regrouping of pre-existing and known elements into new structural arrangements (see Friedmann, 1974, pp. 18–33).
4 R. Griffin described innovation as an organisation's effort to master new products and services or new applications of those existing early on (see Griffin, 1996, p. 476).
5 P. Kotler defined innovation as a good, service or idea that is perceived by someone as new (see Kotler, 1994, p. 322).

2 Efficiency in Energy

CONTENTS

2.1 The Concept and Essence of Efficiency 27
    2.1.1 Traditional Approach 28
    2.1.2 Resource-based Approach 28
    2.1.3 Evaluation of Adopted Strategies 28
2.2 Measurement of Energy Efficiency 30
2.3 Energetics Efficiency and Energy Efficiency 35

2.1 THE CONCEPT AND ESSENCE OF EFFICIENCY

Efficiency is one of the fundamental issues studied and described in economics, management and other academic disciplines. It is also one of the most important criteria for evaluating economic and non-economic activities and is commonly used to represent the state, scope of operation and development opportunities of individual entities.

The concept of efficiency is used very frequently in various contexts and circumstances. However, the widespread use of the concept has not led to an unambiguous definition of efficiency. Many difficulties and doubts therefore arise in relation to the interpretation of the concept of efficiency.

Studies on efficiency indicate that it is a broad phenomenon, reflected in all aspects of socio-economic life. An analysis of the literature on the subject leads one to reflect that efficiency is always a subjective concept, the meaning of which derives from the context of the analysis, or the phenomenon being described (Pasour, 1981, p. 135). Authors dealing with efficiency extensively describe definitions of efficiency in terms of:

- scientific discipline: mathematics, finance, etc. (Glodziński, 2017, pp. 19–21)
- synonyms of effectiveness, e.g. efficiency, effectiveness (Bielawa, 2013, pp. 25–26; Pyszka, 2015, pp. 13–25),
- levels of analysis: local, regional or global (Cormio et al., 2003, pp. 99–130; Jebaraj and Iniyan, 2006, pp. 281–311),
- organisation (Kirchhoff, 1977, pp. 347–355),
- dimensions of efficiency: political, economic-technical, energy, purpose-driven, institutional, etc. (Schlomann et al., 2015, pp. 97–115; Storm, 2018, pp. 302–329; Tobin, 1984, pp. 1–15).

P. Samuelson and W. Nordhaus included efficiency among the key concepts of economics and define it as sets of goods and services provided and delivered with given resources and technology (Samuelson and Nordhaus, 1991, p. 329). Very often in the literature, efficiency is defined in terms of capacity and effectiveness in achieving objectives (Stoner et al., 2001, p. 249; Drucker, 1995, p. 182).

DOI: 10.1201/9781003370352-3

The problem of efficiency can be considered on many levels and from many perspectives, with the most relevant conceptualisations including the traditional (a), resource (b) and evaluation of the organisation's adopted strategies (c).

2.1.1 TRADITIONAL APPROACH

In the traditional approach, efficiency can be considered in terms of the inputs incurred relative to the outputs produced. Efficiency, understood as the reciprocal relationship between inputs and outputs, can be expressed in three ways:

1) as the difference between outputs and inputs (beneficence):
   Outcome > 0 → effects > inputs,
2) as the ratio of effects to inputs (cost-effectiveness):
   Outcome > 1 → inputs < effects,
3) the quotient of the difference between effects and inputs in relation to the inputs incurred:

$$\mathbf{ROI} = \frac{\mathbf{Net\ Profit}}{\mathbf{Cost\ of\ Investment}} 100\%.$$

2.1.2 RESOURCE-BASED APPROACH

The resource approach, on the other hand, considers efficiency in terms of the most efficient allocation of available resources. Efficiency in the functioning of an entity is, in this approach, a consequence of having and making effective use of unique resources and skills.

2.1.3 EVALUATION OF ADOPTED STRATEGIES

From the point of view of evaluating the strategies adopted by an organisation, efficiency can be regarded as an assessment of the performance of the organisation through its activities. In the literature, it is assumed that efficiency is treated as a tool for measuring the effectiveness of an organisation's management and as a way to increase competitiveness by influencing the implementation of strategies and objectives (Skrzypek, 2012, p. 314).

Evaluation of efficiency of a z-target can be considered from two points of view, that is, the actual efficiency (*ex post* evaluation) and the anticipated efficiency (*ex ante* evaluation) (Pszczołowski, 1978, p. 60).

Efficiency assessment therefore always depends on the point from which it is conducted. As we move from the level of the economy as a whole, the organisation, to the organisational unit, the team and ultimately the individual, performance evaluation takes on a more behavioural character, going beyond the prism of productivity. This means that the more performance appraisal is done from the individual's point of view, the more it relates to psychosocial elements.

Efficiency in Energy

Depending on the level of analysis, efficiency can be assessed from different perspectives. It can be considered as static efficiency and dynamic efficiency, where the former focuses on the best allocation of resources and avoidance of waste, and the latter on the prospect of long-term growth.

(Brzozowska et al., 2021b, p. 87)

The static perspective is in line with neoclassical economics linked to the concept of Pareto optimality and the marginalist concept of general equilibrium, assuming the existence of perfect competition. According to this concept, an economy produces efficiently if, given resources, it provides the most desirable set of goods.

When an economy produces at the limit of its productive capacity, it is impossible to increase the production of one good without reducing the production of another good, and an increase in the welfare of an individual cannot be achieved without worsening the situation of another individual. Pareto assumes that the optimal state involves unanimous approval without entailing conflict-inducing efforts to change the distribution of welfare (Zieliński, 2013, pp. 139–140). This results in a subject moving towards static efficiency moving towards achieving the production possibilities curve and oriented towards eliminating losses and wastage of factors of production and optimising their allocation.

Dynamic efficiency, on the other hand, is associated with development taking place from a strategic perspective, the aim of which is to constantly push goals and measures to a higher level. This approach derives from evolutionary economics, which emphasises dynamic phenomena and focuses its attention on observations far from equilibrium. It is therefore characterised by constant change, creativity, strong motivation to act, entrepreneurship and movement generated by highly variable factors. Successful entities are those that can adapt to changes faster and better (Zalega, 2015, p. 161).

Efficiency discourses can not only be time-limited (dependent on comparison) but also be timeless. In the latter case, what matters is a positive change in the input–output ratio, and no matter when in history or in the future it may occur. In practice, however, efficiency research tends to focus on the short term. This is prompted, on the one hand, by the need to define the period of analysis and, on the other, by the desire to demonstrate the effect (Shove, 2018, p. 783).

Efficiency can be considered both from the point of view of individual parties and the economy as a whole.

The efficiency of units can be influenced by a number of factors, some of which include:

- the efficiency of its processes (which can be considered in terms of added value creation defining the increase in value over time as a result of carrying out the intended processes in conjunction with stakeholder expectations);
- the level of fulfilment of formal expectations by individual stakeholders;
- choices made by the units (Chodyński, 2011, p. 261).

When considered at an economy-wide level, it can be assessed from production, allocation, innovation and distribution perspectives. Productive efficiency is achieved when, given supply and demand constraints, production is maximised. Allocative efficiency, on the other hand, is related to the placement of resources in those areas of production that society values most. Innovative efficiency is when the technologies used allow the highest possible efficiency of economic resources. Distributive efficiency, on the other hand, refers to a situation in which no entity tries to change the existing structure of distribution of the produced good.

2.2 MEASUREMENT OF ENERGY EFFICIENCY

The multi-dimensionality and complexity of the concept of efficiency are always associated with difficulties in measuring it. The number of positive and negative factors makes it difficult to select both indicators and yardsticks for measuring the level of efficiency in the energy sector. The assessment of the vector of evaluated changes is also very often controversial. Before examining and measuring efficiency, it must also be clarified that however defined, efficiency is not an end in itself. Efficiency measures must always be accompanied by economic or social objectives that justify the measures taken. Furthermore, it should be kept in mind that excessive exaggeration of efficiency indicators or yardsticks can lead to manipulation and provoke wrong socio-economic decisions. Methods of assessing efficiency in energy or energy effectiveness can therefore be formulated through the prism of the multiple economic, technical, goal-oriented, institutional, political, social, systemic or environmental benefits achieved. The dimensions of efficiency and example measures are presented in Table 2.1.

Depending on the dimension in which efficiency is considered, different metrics can be used. In the case of the economic dimension, efficiency is most often expressed in financial terms or in terms of the effects of an activity, the validity of the use of resources. In the case of the technical dimension, on the other hand, what is most often considered is the extent to which the technical infrastructure is used, expressed in terms of the degree of capacity utilisation. In the case of the goal-oriented aspect, the measure of efficiency is the degree to which objectives

TABLE 2.1
Efficiency Dimensions and Their Characteristics

Efficiency dimension	Characteristics
Economic	Financial and economic evaluation of the results of the activities, validity of the use of physical or natural resources for this purpose
Technical	Evaluation of the use of infrastructure and technical equipment
Goal-oriented	Analysis and evaluation of the extent to which objectives have been achieved
Institutional	An assessment of the support and approval of the activities, which is primarily a function of public perception
Political	Assessment of the achievement of policy objectives
Behavioural	Evaluation of interpersonal relations in the organisation
Systemic	The system of dependencies and links in an organisation, sector, corporate group, etc.
Ecological	The impact of the adopted solutions on the state of the environment

Source: Own elaboration adapted from Pyszka (2015, pp. 21–22).

Efficiency in Energy

are achieved, whereby the objectives can vary widely and depend on the rationale of those setting the objectives. Different characteristics of efficiency also describe the institutional, political, behavioural, systemic or, finally, ecological dimensions. Adopting a particular approach to efficiency may yield opposing observations. For example, the achievement of political or institutional objectives is not always consistent with ecological or economic efficiency. Therefore, the literature on the subject represents the view that does not allow to construct a single and universally accepted measure of efficiency. This is because efficiency assessments always refer to specific spheres of socio-economic life and are carried out by entities that pursue different objectives and have access to limited information. Efficiency assessments are also not free of the individual experiences and beliefs of those analysing the phenomenon under investigation. Indeed, efficiency is analysed not only by the measurable results obtained but also by the subjective feelings of those interpreting them. Efficiency considerations are therefore inevitably burdened by some conception of value, and it is impossible to disengage from the notion that efficiency is somehow more desirable than inefficiency (Blaug, 2000, p. 628).

From the point of view of the main objective undertaken in the book, economic efficiency is particularly relevant. Figure 2.1 presents the procedure for transforming the various efficiency categories into economic efficiency.

There is a wealth of research on the definition, measurement and benefits of implementing energy efficiency measures (Fawcett and Killip, 2019, pp. 1171–1179; Geller et al., 2006, pp. 556–573). In terms of analyses of energy production and use processes, there are differences between looking at energy conversion efficiency problems from the point of view of, for example, thermodynamics versus economic

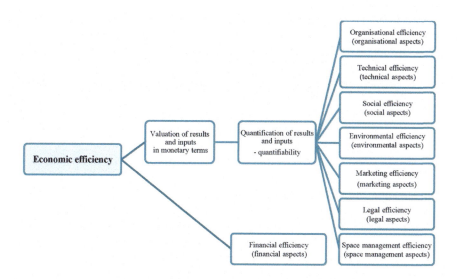

FIGURE 2.1 The procedure for transforming the various efficiency categories into economic efficiency.

Source: Own elaboration adapted from Glodziński (2014, p. 163).

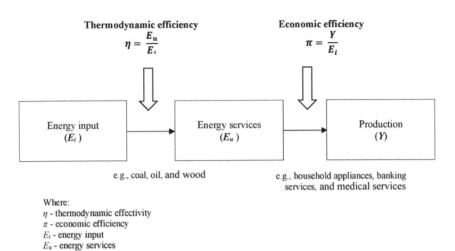

FIGURE 2.2 Energy inputs, energy services and GDP.

Source: Own elaboration adapted from Kander et al. (2013, p. 23).

efficiency. Thermodynamic efficiency can be expressed by the ratio of the utility of energy (or energy services) to the energy source (primary or secondary), whereas economic efficiency can be expressed by the conversion of energy into value-identifiable units (see Figure 2.2).

The conversions presented in Figure 2.2 consequently lead to a specific economic outcome. Energy sources, both primary and secondary, are converted into energy services that are actually demanded. Energy inputs, on the other hand, combine with other factors of production (labour, capital) and together create economic value.

On a macroeconomic scale, changes in energy efficiency can occur as a result of the following (Kander et al., 2013, pp. 22–24):

- Advances in technical knowledge and related innovations, such as the introduction of new equipment or better organisation of work.
- An increase in the level of education and competence of the public or institutions that have an impact on the economy and society.
- Structural changes that lead to changes in production and service provision. The consequence is a reduction in the importance of certain branches of the economy or even their disappearance, and instead an increase in the importance of others or the emergence of new ones. In highly developed countries, industries defined as traditional (e.g. metallurgy and mining) tend to lose importance, while innovative industries play an increasingly important role. As a consequence, this may lead to a situation in which, through the development of industries not dependent on large energy supplies, a significant increase in the share of national income occurs with unchanged or even lower energy consumption. In this case, the energy efficiency of the

Efficiency in Energy

economy will increase, although technical knowledge, innovation or public awareness will not change.

The idea that energy efficiency should be an important part of national (and international) energy policy arose in response to the first oil price crisis in 1973, when reducing energy demand was seen as the solution to achieving greater energy security in many developed countries. Nowadays, it is widely recognised that efficiency in the energy sector should be prioritised (Fawcett and Killip, 2019, pp. 1171–1179; Geller et al., 2006, pp. 556–573). This is determined, among other things, by the fact that efficiency measures in the energy sector are the cheapest and fastest way to achieve large-scale effects.

When taking action to improve energy efficiency, it is useful to find answers to key questions that guide the direction and scope of action:

- Which energy processes are performing optimally, and which ones need to be changed?
- Whether the use of commercially available energy technologies will achieve acceptable economic, environmental and social performance?
- Do we have human resources ready to use and exploit the planned technologies and, if so, what measures should be taken to prepare future users for safe operation?
- At what point to decide to modernise the energy processes?
- What sources of funding are used to finance the investment?

For the determination and estimation of energy efficiency, the ability to quantify and model actual and predicted energy events becomes crucial.

Forecasting in the energy industry is very important because, depending on the scope and purpose of the forecast, it allows rational and efficient planning of energy development, gives a chance to avoid an energy crisis or reduces problems related to the production of excess energy (Morgül Tumbaz and İpek, 2021, pp. 1663–1675). It also becomes particularly important in the process of improving energy efficiency.

Responsive and accurate forecasting in the energy sector helps decision-makers formulate energy development plans and respond to the opportunities and challenges of changing energy demand. Ensuring an adequate energy supply is an important foundation for sustainable socio-economic development. This is particularly relevant for countries that are in the midst of energy transition and economic restructuring, where both energy supply and demand or structural adjustments are key factors determining the effect of the changes being made.

Forecasting involves predicting the future based on trend analysis of current and past data, consisting of three main elements: input variables (historical and current data), forecasting/estimation methods (trend analysis) and output variables (future predictions) (Debnath and Mourshed, 2018, p. 298). The timeframes of the forecast models range from a few hours to sometimes up to 100 years. M. Grubb and his team (Grubb et al., 1993, pp. 397–478) proposed a division into ranges, including short-term forecasts of up to five years, medium term forecasts of 3 to 15 years and long-term forecasts of 10 or more years.

Effective forecasting of energy demand and energy structure helps energy planners formulate development plans and respond to the opportunities and challenges of changing energy demand. Ensuring an adequate energy supply is an important foundation that determines economic development in most countries of the world. Especially for developing countries that are in the process of economic transition and restructuring, energy supply and demand and structural adjustments are key to their transformation. Energy demand estimation influences energy policy, economic development (Hu et al., 2015, pp. 9392–9406) and carbon emission forecasting (Li et al., 2018, p. 2475). It can also serve as a basis for building energy planning models (Rehman et al., 2017, p. 1868). Therefore, accurate energy demand forecasting can be an effective guide for policy makers in energy development planning and policy formulation. It also helps avoid energy supply and demand imbalances and ensures energy security. It also provides an opportunity to plan, organise and implement measures to improve energy efficiency over longer time horizons.

The first energy consumption models were therefore aimed at forecasting energy demand. In the beginning, models were only developed by energy companies and took the form of simple relationships, formulated by trend extrapolation and simple econometric models. In the long term, however, planning proved to be dependent on a number of factors and, as early as in the 1970s, a change in approach to the power system forecasting problem became necessary as a result of economic changes. The starting point for analysis became the end user, who determines demand. The new approach allowed the creation of models with a better fit, but made their formulation more complicated (creating models, from the point of view of the consumer, introduces many more factors into the model of energy consumption that determine the consumption curve).

Predictive models used in the energy industry can be classified in several ways, such as static and dynamic models, one-dimensional and multi-dimensional models, techniques ranging from time series to hybrid models (Suganthi and Samuel, 2012, p. 1224).

Many forecasting models have been proposed to find an effective method that can be applied to practical situations. These techniques mostly rely on complex statistics, artificial intelligence techniques and a large amount of meteorological and topographical data (Bianco et al., 2009, pp. 1413–1421); Karabiber and Xydis, 2020; Rahman et al., 2018, pp. 372–388).

In terms of application, the models used in energy management can be divided into two main groups: the analytical-predictive model group and the predictive model group used in the current study. This distinction is detailed in Table 2.2.

Modelling in the energy industry can also be divided by ways in which data is selected for the model. Two approaches are observed in this case, the so-called 'top-down' and 'bottom-up' modelling. The 'bottom-up' method involves moving from a specific to a general model. The process starts by creating a theory that is a gross simplification of reality. A very simple relationship representing this theory is formulated and the regression equations are estimated from the available data. However, there is a risk of inappropriate model formulation, especially when the number of parameters is limited. Tests of a few basic assumptions, such as Durbin-Watson autocorrelation, are then carried out. Based on the results obtained, the model is corrected and re-estimated (Grycan and Wróblewski, 2016, pp. 168–177).

TABLE 2.2
Breakdown of Forecasting Models in the Energy Economy

Statistical models			Artificial intelligence
Forecasting	Time series models	Naive method	Artificial neural networks
		Exponential smoothing	Fuzzy logic
		Moving average MA	Expert systems
		AR autoregression	Support vectors
		ARMA	
		ARIMA	
		SARIMA	
		ARX, ARMAX, ARIMAX, SARIMAX	
Analytical-prognostic	Econometric models	Static and dynamic	
		(depending on whether the explained variable from previous periods is the explanatory variable (including the concept of consensus modelling)	
		Linear and non-linear	
		(depending on whether the nature of the relationship between the explained variable and the explanatory variables is linear (possibly reducible to linear) or not)	
		Single- or multiple-equation	
		(linear multi-equation models can be divided into: simple, recursive and interdependent equations)	
		Simple (1 explanatory variable)	
		Multivariate (1<number of explanatory variables)	

Source: Own elaboration adapted from Grycan and Wróblewski (2016, pp. 168–177).

2.3 ENERGETICS EFFICIENCY AND ENERGY EFFICIENCY

Efficiency of the energetics is a broad concept, referring to the whole range of functioning of the energy sector. Therefore, when undertaking any energy sector initiative, in addition to considering energetics efficiency, it is necessary to undertake a thorough and professional analysis in terms of financial, organisational, social, environmental, marketing or legal efficiency. Efficiency measures in the energy sector have a significant impact on the socio-economic development of the region and thus on the economic performance of the various entities involved in its operation.

In addition to a general analysis of the efficiency of the entire energy sector or individual solutions, energy efficiency is a significant issue in the energy sector. Energy efficiency is defined most generally as the ratio of outputs, services, goods or energy obtained to energy input. Energy efficiency is of particular importance in EU policy, for example, where references to energy efficiency are made in Directive 2006/32/EC on energy end-use efficiency and energy services (Directive 2003/87/ EC of the European Parliament and of the Council of 13 October 2003 establishing a scheme for greenhouse gas emission allowance trading within the Community and amending Council Directive 96/61/EC). This Directive was introduced to implement the principles of 'cost-effective improvement of energy end-use efficiency'. It defines the terms 'energy efficiency' and 'energy efficiency improvement'. According to the act, energy efficiency is the ratio between outputs, services, goods or energy input and energy efficiency improvement is improvement in energy end-use efficiency as a result of technological, economic or behavioural changes.

The current EU energy efficiency targets contribute significantly to the ambitious EU climate goals. In 2021, European Commission conducted a study on the energy saving potential of energy efficiency measures, which showed that the economic savings potential in 2030 could be as high as 13% of final energy consumption (752 Mtoe) if the EU intensifies its financing and energy efficiency efforts. The technical potential for energy savings could be as high as 19% of final energy consumption (696 Mtoe) (Menge et al., 2021). The EU's experience to date, for example in meeting the 20–20–20 target and the 2030 climate targets set out in the climate law, are clear examples of how energy efficiency measures have enabled economic growth and reduced energy consumption to be achieved simultaneously and have played an important role in meeting the greenhouse gas reduction target. As research confirms, energy efficiency policies have also delivered wider benefits that can be expressed in monetary terms to varying degrees. These include aspects such as reducing fuel poverty, improving health and well-being, increasing business competitiveness and improving energy security (Communication from the Commission to the European Parliament, The Council, The European Economic and Social Committee and the Committee of the Regions, EU 'Save Energy', European Commission, 2022).

Improving energy efficiency is one of the main factors in improving the quality of life and developing entrepreneurship and innovation. The process of taking action to reduce energy losses is a widely socially accepted measure of sustainable development (An Integrated Industrial Policy for the Globalisation Era, Communication from the Commission to the European Parliament, The Council, The European Economic and Social Committee and the Committee of the Regions, COM/2010/0614 final, 2010, p. 22). Energy efficiency improvements can take place in all fields of socio-economic activity, especially in:

- industrial production,
- agriculture,
- buildings (public buildings, businesses, especially in housing),
- the way citizens live and consume,
- energy infrastructure,
- transport.

The opportunities for improving energy efficiency in industry and agriculture are vast. These include innovations leading to energy efficiency gains in industrial processes (relating to the scope of production), multi-sectoral measures (e.g. energy efficiency labelling systems, product energy labels or smart metering systems) and horizontal measures (promoting energy efficiency improvements and

Efficiency in Energy

energy efficiency measures) (Skoczkowski and Bielecki, 2016, pp. 9–14). In the case of agriculture, the opportunities for energy efficiency are also enormous, with heating, dehumidification, lighting, ventilation and cooling being among the main areas in which energy saving can be sought. Energy efficiency gains in agriculture can therefore mainly be achieved in the process of energy use and in increasing the use of renewable energy. The high retail prices of fuels and electricity also contribute to their conservation and the search for their substitution by other cheaper energy carriers available in agriculture. Significant energy effects in case of agriculture will also be obtained as a result of the decreasing number of small farms, where it is difficult to separate the energy consumption incurred directly for agricultural production and for the social and subsistence activities of the farmer's family (Wójcicki, 2006, p. 41).

The energy efficiency potential for existing buildings is very high. For example, in Poland, as many as 70% of single-family homes are heated by coal (approximately 3.5 million). The vast majority of these installations (approx. 3 million) are coal boilers, technologically obsolete appliances with low efficiency and significant low emissions. Furthermore, it should be kept in mind that it is also very common for new buildings to install coal boilers (Pytliński, 2013, p. 11). Savings made in energy consumption in buildings can therefore become an important factor in increasing energy efficiency. In this aspect, the requirements for the thermal protection of buildings are of particular importance, which have been and, it seems, will continue to become more and more rigorous over the decades. Newly constructed buildings are now required to meet at least the standard of an energy-saving building.

Table 2.3 shows the age structure of the housing stock in Poland, together with estimates of its specific demand for non-renewable primary energy (PE) and final energy of that stock (FE).

TABLE 2.3
Age Structure of the Housing Stock in Poland Built Before 2002 and Its Unit Energy Demand Indicators

Construction period	Buildings	PE	FE
Years	Thousands	kWh/(m²-year)	kWh/(m²-year)
Before 1918	404.7	> 350	> 300
1918–1944	803.9	300–350	260–300
1945–1970	1363.9	250–300	220–260
1971–1978	659.8	210–250	190–220
1979–1988	754.0	160–210	140–190
1989–2002	670.9	140–180	125–160

Source: Own elaboration adapted from Long-term Renovation Strategy. Supporting the Renovation of the National Construction Resource, Warsaw, 2022. Annex to Resolution No. 23/2022 of the Council of Ministers of February 9, 2022, p. 21.

Note: PE is an indicator for annual non-renewable primary energy demand per unit area of temperature-controlled space, expressed in kWh/(m²-yr), and FE is an indicator for annual final energy demand per unit area of temperature-controlled space, expressed in kWh/(m²-yr).

As it is clear from the analysis of the data in Table 2.3, the oldest buildings, of which the largest number of them in Poland, were built to the worst energy standards. On the one hand, this means that there is great potential for improving the energy efficiency of these buildings, but on the other hand, it must be kept in mind that these buildings are very often inhabited by the poorest households, who cannot afford to make thermo-modernisation investments. The low energy standards of buildings, combined very often with the poor technical condition of the energy infrastructure, lead to many social problems, including energy poverty.

Energy efficiency in buildings can be increased by:

- building new, energy-efficient buildings,
- reconstruction or renovation of existing buildings, along with thermo-modernisation works,
- introducing intelligent energy management systems in buildings.

Estimates have been made that cost-effective thermo-modernisation of residential buildings potentially achieve a reduction in dust emissions of around 89,000 tonnes per year, which is around a quarter of total dust emissions in Poland (Long-term Renovation Strategy. Supporting the Renovation of the National Construction Resource, Warsaw, 2022. Annex to Resolution No. 23/2022 of the Council of Ministers of February 9, 2022, p. 54).

Another issue with a significant impact on increasing energy efficiency is the degree of public acceptance of the various methods and ways of achieving it. In this aspect, society's ability to save energy and its ability to use new technologies are important. Innovative attitudes also play a huge role in this regard and are crucial to the acceptance of new solutions, including those based on renewable energy sources. In order to successfully involve the public in actively improving energy efficiency, it is essential to ensure that citizens have the necessary standard of living and financial capacity to commit capital to development projects. A lack of adequate capital or the ability to raise it can prevent the implementation of energy efficiency improvement processes and, in many cases, have the opposite effect. Households without adequate financial resources or opportunities for assistance and support may seek to secure their own energy needs using the simplest and cheapest methods, incompatible with efficient use and environmental protection.

A fundamental influence on the possibilities to improve energy efficiency is the way citizens live and consume. On the one hand, the public is the primary group consuming goods and services in the economy, causing businesses to adapt their products and services and the way they provide them to the needs and expectations of their customers. On the other hand, citizens own the infrastructure, run and work in businesses, pursue investments and make political choices.

The public's awareness of the importance of and opportunities for energy efficiency improvements is critically influenced by education. According to extensive research, education at every stage of life plays an important role in shaping attitudes on how to approach energy efficiency improvement problems (see Alvarado et al., 2021; Pablo-Romero and Sánchez-Braza, 2015, pp. 420–429; Solarin, 2020, pp. 254–265).

Efficiency in Energy

The technical state of the energy infrastructure is of particular importance for improving energy efficiency. Opportunities in this matter should be seen in increasing the efficiency of energy generation, transmission and distribution. Significant opportunities for increasing energy efficiency can also be found in transport, where the primary areas of energy efficiency improvement include (Motowidlak, 2014, p. 4508):

- improving the energy efficiency of vehicles in all modes of transport,
- expansion and improvement of transport infrastructure to enable more efficient organisation of transport (mainly through the development of public transport),
- optimising the operation of multimodal logistics chains (through the widespread use of resource-efficient measures),
- maximum efficient use of transport modes and infrastructure through better traffic management systems and advanced logistics and market measures.

A major consumer of energy resources is transport, and it is therefore in this sector that significant opportunities and potentials for improving energy efficiency are sought. Efficient transport, understood as meeting the demand for movement services while minimising the resources used for this purpose, can be improved in terms of its energy efficiency through, among other things (Woodcock et al., 2007, pp. 1084–1086):

- measures to reduce the transport intensity, that is, the demand for passenger and freight transport;
- measures to improve vehicle technology;
- measures to improve the organisation of transport services;
- the development of transport infrastructure to ensure traffic flow and improve the competitiveness of low-emission and environmentally friendly transport,
- promoting appropriate transport behaviour patterns.

The main instruments to support interest in implementing energy efficiency projects are undoubtedly legal and administrative regulations and all types of financial incentives available for this purpose (subsidies, tax deductions, etc.). In addition to the potential of the aforementioned instruments to influence the main groups of entities responsible for energy consumption, a number of methods and ways to support the energy efficiency improvement processes should also be considered. These include:

a) ESCO formula, energy performance contracting (EPC) and energy delivery contracting (EDC)

Energy efficiency measures can be undertaken with the participation and cooperation of ESCOs. ESCOs offer energy services to improve the energy efficiency of their customers and share the benefits of reduced energy costs. They make investments using their own technical and organisational solutions and provide financing. ESCOs are growing rapidly worldwide, especially in the EU, where they now offer a wide range of services including (Bertoldi and Boza-Kiss, 2017, pp. 345–355; Hansen et al., 2009, p. 8):

- feasibility studies,
- energy audits,
- acquisition and installation of equipment,
- infrastructure and energy management,
- energy supply,
- risk management,
- automated surveying,
- ensuring adequate indoor air quality,
- training and provision of information,
- support of consultancy on sustainable development and environmental protection,
- measurement and verification of energy savings,
- guarantee of achieving the intended results.

ESCO projects can be implemented in two ways (Gulczynski, 2009, pp. 175–182):

- ESCOs are involved from the beginning of a project through to its post-completion management (BOO),
- ESCOs prepare, set up and manage projects within contractual periods and then hand over the infrastructure to the owners (BOOT) on completion.

In the implementation of projects through the ESCO procedure, the energy performance contracting (EPC) and the energy delivery contracting (EDC) are of particular importance. An EPC is a type of contract between the client and the provider of energy efficiency measures (ESCO). It can set out both the technical and financial terms, as well as how the energy savings achieved will be measured and the terms of the guarantee for achieving the savings. It is worth noting that if an EPC contract is not implemented, the supplier of the energy-efficient equipment is only responsible for the quality of the equipment and not for the energy savings achieved. Once an EPC contract is adopted, on the other hand, the service provider becomes responsible not only for the equipment but also for the achieved energy savings (Carbonara and Pellegrino, 2018, pp. 1064–1066; Shang et al., 2015, pp. 60–71).

The services offered by ESCOs differ in terms of how they are financed and how they share the risks and returns from the implemented investment (Lee et al., 2015, pp. 116–127).

An EDC is a contract which sets out the conditions for the operation, construction or modernisation of energy sources (heating and electricity) undertaken at the contractor's own risk. For a contract to be concluded, its scope must cover a sufficiently long period of time during which the energy optimisation process will enable economic and environmental benefits to be achieved in a way that is satisfactory for the entities (Handbook Addressed to Public Sector Entities, KAPE, Warsaw, 2012, p. 24).

Depending on the preferences and decision-making considerations, bearing in mind the differences between EPC and EDC contracts presented in Table 2.4, a

Efficiency in Energy

TABLE 2.4

Comparison of EPC and EDC Contracts

Specification	EPC	EDC
Application	Implementation of investments in the entire area of energy use (supply and demand)	Investment in new energy sources, in the replacement, modernisation, construction and/or expansion of existing energy sources
Contracted services	Financing, planning, installation and maintenance of specific solutions to reduce energy consumption	Financing, planning, installation and operation of energy sources (media supply)
Contract rate (funding)	User charges as remuneration to the contractor for the energy and operating cost savings achieved	Energy supply charge (heating, electricity, cooling)
Advantages	The technological advantage of the contractor makes it possible to achieve high guaranteed savings in energy costs over the duration of the contract and possibly, through other financial incentives, attractive bonuses.	The contractor's market advantage offers the possibility of favourable conditions for the purchase of supplied energy; investments in new installations entail increase in efficiency
Features or principles of contracts	Objective of the contract: guaranteed savings in energy consumption and operating costs: - risk sharing - duration of the contract - distribution of the savings - determination of reference energy costs	Objective of the contract: supply of heating, electricity or cooling (media supply): - risk sharing - duration of the contract - determination of media demand - determination of supply limit

Source: Own elaboration adapted from Handbook Addressed to Public Sector Entities (2012, p. 25).

decision may be made on how to implement projects in an ESCO procedure. The decision should be analysed in an appropriate manner, considering all the costs and benefits that will be generated as a result of the agreement (both for the ESCO and for the entities using their services).

ESCO projects are of great interest to the public sector. The public sector's main barriers to energy efficiency projects include budgetary constraints, lack of knowledge and lack of staff to prepare and implement projects (Lee et al., 2003, pp. 651–657). The ESCO formula, by bringing private sector partners into energy efficiency projects, allows these barriers to be addressed to a significant extent.

b) Constructing an efficient energy pricing system

In order to meet the stated objectives of improving energy efficiency, it is necessary to build a correct pricing policy for the costs incurred by energy consumers. Prices should include the costs of the energy itself, fixed costs (e.g. distribution costs, which are incurred irrespective of the amount of energy consumed) and external costs, where possible. An adequately calculated energy price forces consumers to choose the most efficient ways of obtaining and using energy. It should be kept in mind here that lowered energy prices lead to a situation where energy-saving investments become unattractive. However, excessively high energy prices exacerbate the operating conditions of economic entities or the quality of life and, as a result, may contribute to the search for opportunities to obtain energy in ways that are inefficient for society or the economy (Gulczyński, 2009, pp. 175–182).

c) Energy efficiency agencies and information points

Energy efficiency improvement measures need to be supported by entities who have the necessary knowledge and skills to provide information on and support for possible measures. Due to the dynamic pace of change in organisational, financial, technical or legal terms, most of the entities who can undertake energy efficiency measures do not have the appropriate knowledge or the capacity to acquire it quickly. It is therefore becoming necessary to create specialised agencies (both at national and regional levels) and information points, which should become an important part of the efforts to disseminate knowledge on energy efficiency.

d) Energy audits and labelling

A key aspect of efficient energy management in the building industry is to have the as complete as possible knowledge of the energy characteristics of the building. Energy audits are an important tool for improving the energy efficiency of buildings. The main objective of energy audits is to determine the scope and technical, organisational and economic parameters of a thermo-modernisation project and to develop a recommendation for an optimal solution, especially from the point of view of investment implementation costs and energy savings (Życzyńska, 2013, p. 108).

A specific solution to support energy efficiency is energy labelling of products. By introducing such solution, comprehensive information on the technical data and energy efficiency of a specific appliance can be provided to the customer (Gulczynski, 2009, pp. 178–179).

3 Costs in the Energy Sector

CONTENTS

3.1 Characteristics of Costs in the Energy Industry and Their Classification 43
3.2 Factors Determining Costs in the Energy Industry 47
    3.2.1 Energy Externalities 47
    3.2.2 External Costs Versus Externalities 50
3.3 Internalisation of External Costs in the Energy Industry 52
    3.3.1 Pigou Tax 53
    3.3.2 The Coase Theorem 54
    3.3.3 Methodology for Determination of External Costs 55
3.4 Instruments for Internalising Costs 59
    3.4.1 Environmental Taxes and Charges 59
    3.4.2 Voluntary Agreements 62
    3.4.3 Ecological Compensation 62
    3.4.4 Fiscal Reform 62
    3.4.5 Deposit Fees 64
    3.4.6 Financial Penalties 64
    3.4.7 Direct Regulation 64
    3.4.8 Instruments Based on Market Transactions 65
    3.4.9 Subsidies 67

3.1 CHARACTERISTICS OF COSTS IN THE ENERGY INDUSTRY AND THEIR CLASSIFICATION

Cost is a fundamental criterion in making decisions about energy systems and related changes. However, there is no one-size-fits-all approach to how these costs should be assessed and analysed in business practice.

The implementation of the chosen energy mix in the economy should be based on a full analysis of the capital expenditure, operating costs and project completion. Besides, it is also necessary to evaluate the changes that a given solution adopted in the energy sector generates in the economy, society or the environment. Therefore, any decision taken in the energy sector should take into account all broadly defined costs (including those incurred by third parties). Therefore, aspects such as the following cannot be ignored in the assessment (Tiwari and Mishra, 2012, p. 483):

- investment expenditure,
- the costs of operating and maintaining energy installations,
- fuel/energy feedstock costs,

DOI: 10.1201/9781003370352-4

- replacement costs necessary to ensure the proper functioning of the infrastructure,
- costs associated with controlling and managing emissions,
- waste management costs,
- health costs,
- environmental costs,
- lifetime of the installation,
- the cost of securing the system,
- public and national security costs,
- other external costs.

Before analysing the costs of individual power generation solutions and technologies, it is important to consider the many characteristics of individual power technologies that influence final decisions. Key features in this respect undoubtedly include:

- the size of the unit, directly influencing its applicability in a given sector of economy (region),
- design and implementation period,
- access to components and parts,
- supply chains,
- cost of capital,
- operating costs,
- fuel costs,
- CO_2 costs,
- regulatory risk.

Due to the risks involved in the market, those making investment decisions currently prefer technologies with a short payback period and a short construction period. However, it should be kept in mind that the power sector will be increasingly affected by environmental policy, including the need to reduce CO_2 emissions. It is forcing a huge change in the development of the manufacturing sector, especially in the directions associated with clean coal technologies and nuclear energy, and especially in favour of renewable energy sources (including decentralised systems).

In addition, the choice of future technologies depends on many other factors.

A summary of the key features of manufacturing technologies is shown in Table 3.1.

Analysis of the data presented in Table 3.1 clearly shows the main differences between renewable energy sources and conventional technologies. These include:

- unit size,
- the time needed to design and implement the chosen solution.

The indisputable benefits of RES technology, however, are the absence of fuel purchase costs or no or low harmful emissions into the atmosphere.

Costs in renewable energy are no longer characterised by high CAPEX costs, which, in the case of lower operating costs or project completion costs compared to conventional sources, make renewable energy sources increasingly attractive from

TABLE 3.1

Summary of Energy Technology Features

Technology	Unit size	Design and implementation period	Capital costs/kW	Operating costs	Fuel costs	CO_2	Regulatory risk
CCGT	Average	Short	Low	Low	High	Average	Low
Fossil-fuel power stations	Large	Long	High	Average	Average	High	High
Nuclear power plants	Very large	Long	High	Average	Low	No	High
Hydroelectric power stations	Large	Long	Very high	Very low	No	No	High
Wind power plants	Small	Short	High	Very low	No	No	Average
Power plants with piston engines	Small	Very short	Low	Low	High	Average	Average
Fuel cells	Small	Very short	Very high	Average	High	Average	Low
Photovoltaic cells	Very small	Very short	Very high	Very low	No	No	Low

Source: Own elaboration adapted from Szczerbowski (2014, p. 11).

the point of view of capital expenditure. Examples of investment costs for energy generation technologies are presented in Table 3.2.

As can be seen from the estimates presented in Table 3.2, investment in fossil fuel-fired generation units will increase in the coming decades. This will particularly be the case in highly developed countries. This will be determined in particular by increasing environmental requirements as well as energy and national security issues. A significant improvement in the efficiency of renewable energy is therefore to be expected, as expressed in a decrease in0 CAPEX. Some renewable energy sources are already more attractive than conventional energy, as a result of continuous improvements in technology and the increasing long-term operating costs of new coal- or gas-fired power plants (Sens et al., 2022, pp. 525–537).

An important conclusion from the investment cost comparison is that key renewable technologies are becoming increasingly competitive with new fossil fuel (coal or gas) power plants.

However, it should be kept in mind that the costs incurred directly in the energy sector very often do not fully reflect the degree of use of scarce resources and the effects that arise as a result of burning fossil fuels or the risks associated with the operation of nuclear power industry. Particularly relevant is also the increasing impact of the geostrategic factor on the energy sector, related to military conflicts and the problem of the risk of hybrid war (Księżopolski et al., 2021, p. 331).

In addition to the aforementioned determinants, the energy models, on the basis of which the energy systems operate, have a significant impact on costs in the energy

TABLE 3.2
Unit Investment Costs for Energy Generation Technologies (EUR/kWe)

Specification	2020	2030	2050
Gas turbine	510	530	570
Hard coal-fired unit	1,575	1,625	1,725
Lignite-fired unit	1,825	1,875	1,975
Nuclear power plant	5,000	5,000	5,000
Onshore wind farm	1,400	1,300	1,200
Offshore wind farm	2,800	2,500	1,950
Photovoltaics	800	600	475
Run-of-river hydroelectric power plants	3,600	3,600	3,600
Biogas	3,200	3,100	2,900
Biomass/waste	3,100	3,100	3,100

Source: Own elaboration adapted from Ecke et al. (2017, p. 29).

sector. From the point of view of assessing the efficiency of civic energy development, the high-efficiency organisation model proposed by Barczak (2016, p. 143) is of interest. It uses the alignment of the main elements in a way that allows high efficiency to be achieved in a positive context. The main elements of the model include strategy, culture, people, structure, tasks and systems. Table 3.3 presents the highly efficient organisation model along with organisational elements and its features.

Given the research scope, models for forecasting energy system development can be divided into (Karkour et al., 2020, p. 1; Kudełko, 2005, pp. 250–252; Pfenninger et al., 2014, pp. 74–86):

- energy system models (showing the characteristics of the energy market, the entities in the market and the ways in which energy carriers are obtained, processed and used),
- energy-economics models (representing the relationships and interdependencies between the energy system and the economy),
- integrated energy-economics-environmental models (consisting of several specialised and complementary models mapping the technological, economic and environmental relationships).

The feedbacks that occur between the energy system and the rest of the economy mean that, when constructing energy models, the sector's links to the rest of the economy cannot be abstracted. Therefore, the construction of energy models should consist of four key elements:

- energy, showing the interdependencies within the energy system (issues relating to energy production, the extent of energy exports/imports, transport of energy carriers, emissions or energy security),
- political (taking into account political considerations, including those related to the public sector's provision of public, health, social or national security),

TABLE 3.3
Highly Efficient Organisation Model

Organisational element	Features of the element
Strategy	A long-term strategy of action with clear and achievable objectives based on the actual development needs and aspirations of those who adopt them
Culture	Relationships based on participation, trust, cooperation and openness to new ideas
People	Committed, understanding and identifying with the organisation's operational objectives
Structure	Flexible and able to quickly accept changes and challenges from the environment
Tasks	Sharing information and knowledge, continuously improving new processes, services and products, creating value for stakeholders, guided by respect for the common good
Systems	Fair and clear rules on remuneration and motivation, open communication.

Source: Own elaboration adapted from Barczak (2016, p. 143).

- economic, characterising the other elements of the economy (final demand, employment, income, prices, business cycles, economic growth, monetary policy) and their impact on energy,
- social (taking into account the needs and aspirations of the public for participation in the energy system in terms of decision-making, equity and ownership).

This approach leads to a situation where the distinguishing feature of the energy models being prepared in modern economies and civil societies is the network character of organisation and cooperation. In particular, economic and social networks are becoming important. These networks intermingle, interact intensively and create specific conditions for competition, cooperation and the influence of the organisation on the environment and vice versa (Mikuła, 2006, p. 26). Networking is especially important in case of civic energy, which, due to its local character, which is based on dispersed renewable energy sources, is particularly significant.

3.2 FACTORS DETERMINING COSTS IN THE ENERGY INDUSTRY

3.2.1 ENERGY EXTERNALITIES

The existence of externalities is related to *market failure* and their occurrence accompanies production and consumption processes. R. Coase defines externalities as the positive and negative effects suffered by those who are not owners of the good used, participants in the production or exchange process, and who have not agreed to face these effects through a voluntary agreement (Coase, 1960, p. 1–14).

Externalities can be divided into public and private:

- Public externalities are consumed by all entities in an area. Consumption by one entity does not affect the consumption levels of other entities, meaning that consumption of an effect by one entity does not reduce the availability or utility of that effect for other entities;
- Private externalities are affected by attrition. The consumption of an effect by one actor implies an attrition of consumption of that effect by other actors.

The energy sector is a typical area where market disturbances occur. It is increasingly accepted in the literature that energy development planning should be done on the basis of full costing, taking into account both the direct costs of building and operating energy facilities and the monetary value of the externalities they cause. Introducing externalities into socio-economic analyses requires identifying and understanding the market forces that determine the energy mix used (Chow et al., 2003, p. 1530).

Externalities can affect different areas and be classified as local, regional or global. In discussions of externalities, a distinction is often made between technological and monetary externalities. Technological externalities refer to a situation where the actions of one entity affect the utility or competitiveness of another entity. Monetary effects refer to a situation in which the actions of an entity or entities affect the price level (Greenwald and Stiglitz, 1986, pp. 229–264). The main externalities caused by the energy sector in the main areas of influence are:

- Human health: reduction in human life expectancy, hospital treatment related to cardiovascular and respiratory diseases, etc.,
- Material damage: corrosion of metals, deterioration of building facades, loss of materials, etc.,
- Crop damage: reduced productivity of agricultural crops, forestry losses, need for increased nitration of soils,
- Loss of biodiversity: biological imbalance,
- Political and military conflicts: energy dependence of countries or regions and the use of energy resources or the energy sector as a tool for political or economic blackmail,
- Global warming: threat to life, floods, loss of agriculture, tourism, etc.

In case of the impact of energy on human health, the externalities arising are exclusively negative. Through its activities, the energy industry directly contributes to the deterioration of human health. This is expressed in increased morbidity or mortality, which directly translates into economic problems (e.g. costs of absence from work, provision of medical care), and in social issues, related to problems brought about by diseases occurring in the family or the environment. Of course, it is not possible to clearly separate the area of influence of a specific externality. This is because they affect different areas with different intensity and at different times, with changes in one area affecting the functioning of others.

Costs in the Energy Sector

One of the main reasons for economics' interest in the issue of externalities has been the problem of environmental pollution. The adverse environmental effects of burning fossil fuels also suggest that current patterns of energy use are unsustainable in the long term. Action to eliminate negative externalities seeks to remove the discrepancy between private and social costs. This can be achieved through direct regulation, such as taxation, the introduction of production or service delivery standards.

Figure 3.1 presents an example of the impact of negative externalities on the welfare changes of the three main market participants: producers, consumers and the state (market regulator). It is assumed that negative effects increase in proportion to the volume of production and only affect the country in which they arise. The costs associated with the creation of negative externalities resulting from the production process are faced by society. Thus, the private marginal cost (PMC) is below the social marginal cost (SMC) at each level of production, and the discrepancy between SMC and PMC increases proportionally as production increases (Hitiris, 2003, p. 318).

In the absence of any policy regulating the actions of producers, they set output according to the private supply curve and the competitive market reaches equilibrium (with price P_0 and output Q_0). In such a situation, the welfare level is as follows (Hitiris, 2003, p. 318):

- consumers = x + a + b + c,
- producers = h + k + e + f + g,
- negative effects = − (k + f + b + g + c + d),
- net effect = (x + a + e + h) − d.

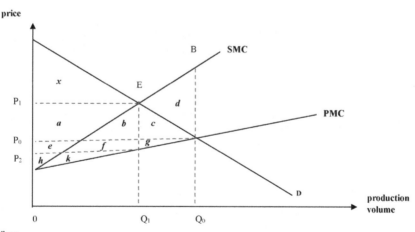

Where:
D = demand; PMC = private marginal cost; SMC = social marginal cost.

FIGURE 3.1 The impact of negative externalities on the welfare changes of the three primary market participants: producers, consumers and the state.

Source: Own elaboration adapted from Hitiris (2003, p. 318).

50 Management of Civic Energy and the Green Transformation

With the implementation of policies to counteract negative externalities through optimal taxation, the equilibrium state is reached at P_I and with output Q_I. In such situation, the welfare level is as follows (Hitiris, 2003, p. 318):

- consumers = x,
- producers = h,
- row = a + b + e + f,
- negative effects = − (k + f + b),
- net effect = (x + a + e + h) − k.

With the implementation of policies to counteract negative externalities by regulating supply, a state of equilibrium is reached at the price point P_I and with production Q_I. In such situation, the welfare level is as follows (Hitiris, 2003, p. 318):

- consumers = x,
- producers = h + e + a + b,
- negative effects = − (k + f + b),
- Net effect = (x + a + e + h) − k.

When optimal taxation or supply regulation policies are implemented, the overall effect is identical and 'better' than in the absence of any policy, but the distributional effects of the two policies between consumers, producers and the government are different (Hitiris, 2003, p. 318).

3.2.2 EXTERNAL COSTS VERSUS EXTERNALITIES

In the extensive literature on market failures and the emergence of externalities that subsequently contribute to the generation of external costs, there is no clear conceptual distinction between external effects (i.e. externalities) and external costs. As a rule, externalities are defined broadly and include measurable and non-measurable phenomena. External costs, on the other hand, are presented in measurable terms which, through internalisation, can be included in production or consumption costs. Thus, the process of converting adverse externalities into monetary units makes it possible to calculate the amount of external costs. External costs arise when two conditions are met simultaneously (Pearce and Turner, 1990, pp. 100–101):

- the activities of one entity affect the welfare of the other (leading to its deterioration),
- deterioration in welfare is not compensated for.

In economic practice, very often the expenses related to energy investments or energy costs do not reflect the total costs of fossil fuels for individuals or for a society as a whole. Despite the serious effects on health and the environment, known as externalities, the hidden costs of fossil fuels are not represented in their market price. Figure 3.2 shows the relationship between distorted prices and real prices in the energy industry. Distorted prices include hidden costs that are indirectly faced

Costs in the Energy Sector

by society. These are external costs, which when introduced into the economic calculation, make market prices more realistic. Real prices therefore contain external costs that are real and fully 'external' because they are kept by third parties and future generations, rather than directly by their generators or consumers. In the case of energy, it is not in society's interest to eliminate external costs, as this would be too costly and suboptimal.

A key issue in terms of analysing the direction of conventional, nuclear or renewable energy development is the different approach to the problem of economic risk and the possible pace of change. For both conventional and nuclear power, the process of opting out is lengthy. They also tend to be state-owned and state-controlled, which results in a situation where, due to investment, social or legal considerations, any attempt at change is met with resistance from various interest groups. In the case of renewable energy, the situation is quite different. Renewable energy is controlled by many individuals, dispersed entities who take part in the market game. Due to the scale and nature of the investments, the eventual exit from the market and 'abandonment' of energy production is easy and does not result in significant structural changes. Any political risk is therefore negligible, and the costs are faced by consumers and taxpayers virtually in real time.

When considering the external costs of energy, one should also take into account the problem of security of supply and infrastructure failure. At the same time, this problem can be considered very broadly and on different levels. One example is the global scandal involving the Enron Corporation, which failed to maintain and upgrade its energy infrastructure, which, together with crimes related to the falsification of financial records, led many thousands of entities to lose billions of dollars (Li, 2010, pp. 37–39). On a local and regional scale, this can include problems with outdated or inadequate infrastructure. It is worth mentioning here the transmission networks, which are very often in poor technical condition and in need of urgent renovation.

When analysing the issue of external costs in fossil energy and highlighting the external benefits in renewable energy, one cannot forget about the external costs that also arise in renewable energy. Although renewable energy is classified as a green, zero-carbon energy source, it can generate many negative external costs. These are generally costs associated with the production process of RES installations or costs associated with the need to operate, maintain or service the installations. Disposal costs, which can in many cases be significant, must also be taken into account. Furthermore, it is important to bear in mind that RES use technologies that have very often been in use for a short period of time, so the negative effects they generate cannot be clearly identified (e.g. the generation of different effects by wind turbines and their impact on humans or the environment remains an open question).

External costs in energy analyses should be considered primarily because of the possibilities to choose the most efficient type of technological solutions. Available analyses indicate that the largest external costs related to environmental pollution and human health impacts dominate mainly in fossil energy (Renewable Energy, 2012, p. 147). External costs related to climate change dominate especially for fossil energy if not equipped with CCS. It should also be kept in mind that the dominance of coal and oil in the global energy system relies heavily on massive public subsidies (Fücks, 2016, p. 329).

Solving the problem of the negative externalities (costs) is possible through the use of methods of their internalisation, understood as ways leading to the transformation of external disadvantages into internal costs of functioning of economic entities (Owen, 2004, pp. 127–154). Ways of internalising external costs determine the scope and manner of applying specific financial or economic incentives in order for investors to choose the most efficient solutions from socio-economic and environmental perspectives.

3.3 INTERNALISATION OF EXTERNAL COSTS IN THE ENERGY INDUSTRY

The definition of external costs in the energy industry refers to all negative effects associated with the generation, transmission and consumption of electricity, heat energy at all stages of the technical process, which include (Friedrich and Voss, 1993, pp. 114–115):

- construction or decommissioning of energy infrastructure,
- extraction or transport of energy resources,
- emissions from the production and subsequent consumption of energy.

In this sense, external costs are defined for the entire process chain, which means that they are always specific to individual energy fuels. Failure to take external costs into account in the energy industry leads to overexploitation of production factors that could be used in other alternative areas of the economy. The lack of efficient use of energy raw materials manifested in a negative impact on the environment results in a reduction in social welfare.

Internalisation of costs is commonly defined as the process of integrating negative externalities relating to environmental degradation and the use of non-renewable resources in household and business budgets through economic instruments, including fiscal measures or incentives (Glossary of Environment Statistics, Studies in Methods, Series F, No. 67, 1997).

Internalisation is therefore a way of bringing the private and social cost of production closer together and should address the following issues:

- determine the scale of acceptable use of the environment;
- determine how this scope is to be distributed among those who use the environment.

The dynamic changes taking place in the energy system are fostering a structural redevelopment of the energy industry. There is also growing public awareness of the need to internalise the external costs of energy as a universal, simple and transparent solution. The basic problem that arises is the choice of how to achieve an optimal level of production. This task can be accomplished in two ways:

- by state intervention,
- by defining proprietorships.

3.3.1 Pigou Tax

Government intervention in the market is generally justified in situations where the market fails to provide an efficient allocation of resources, that is, where the market becomes unreliable (market failure) (Masur and Posner, 2015, p. 5). A.C. Pigou assumed that the compensation of costs resulting from the existence of externalities should take place administratively and be regulated by a special tax.1 A Pigou tax is defined as a tax levied on the entity causing environmental pollution (environmental damage) (Bergmann, 2009, p. 131).

The application of the Pigou tax makes it possible to correct the misallocation resulting from the omission of externalities. If one considers a situation where there is a competitive market in which negative externalities arise, the consequence will be an inefficiently high level of production from the point of view of the society. Such situation is presented in Figure 3.2, where negative externalities give rise to a marginal damage (MD) that causes the social marginal cost (SMC) to exceed the private marginal cost (PMC). The difference between the social marginal cost curve and the private marginal cost curve is the amount of marginal damage caused by externalities.

In order to equalise private marginal cost with social marginal cost, corrective taxes must be introduced (see Figure 3.2). A tax of t equal to the marginal damage shifts the supply curve (from S to $S+t$), resulting in a situation where it intersects with the demand curve at the optimal level from the point of view of social marginal cost.

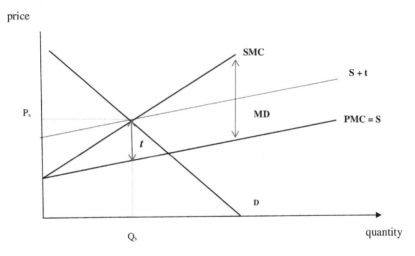

Where:
D = Demand,
S = Supply,
MD = Marginal Damage,
PMC = Private Marginal Cost,
SMC = Social Marginal Cost.

FIGURE 3.2 Pigou tax.

Source: Own elaboration.

Pigou's proposed tax allows for the integration of social objectives with those of economic entities. Nevertheless, in economic practice, the implementation of the Pigou tax faces a number of problems. First, it is in fact difficult (or practically impossible) to estimate the exact level of the Pigou tax, which is determined by the need for accurate information on marginal damage. In addition, the concept of the tax assumes that the optimum level of pollution is achieved by imposing a tax at a level that forces economic operators to reduce production volumes. This may lead to an increase in the price of goods or cause distributional effects (e.g. a change in the proportion of profit distribution). It may also happen that, due to the imposed tax, manufacturers who cannot cope with the additional fiscal burden will be eliminated from the market (Pomaskow, 2016, pp. 100–102). These reasons make Pigou taxes almost non-existent in economic practice.

3.3.2 THE COASE THEOREM

An important point in considering externalities is the theorem formulated by R.H. Coase (Coase, 1960, pp. 1–44). He assumed that there is no need to interfere with the market to take externalities into account when there are clearly defined property rights associated with the disputed situation.

In his work, R.H. Coase showed that the conclusions formulated by Pigou (and his followers) may or may not lead to efficient allocation (Jastrzebski and Mroczek, 2014, p. 158). Thus, despite the impact of externalities in production or consumption, the market mechanism may lead to some kind of optimum in the Pareto sense, so there may be solutions based on private agreements that can eliminate externalities. The key remains the transaction costs of these arrangements, which may be so large that it makes much more sense to rely on legislation or state action to eliminate market failures (Blaug, 2000, p. 619).

R.H. Coase has also shown that the view represented by A.C. Pigou, according to which only the public authority has the instruments to reduce and ultimately also eliminate negative externalities, is completely unfounded. Indeed, not only is it possible to internalise negative externalities by means of the free market, but there are also many examples of the public sector sanctioning the occurrence of negative effects, aiming to perpetuate them rather than eliminate them (Golecki, 2011, p. 166).

The solution of externalities by means of Coase's theorem is very often criticised in the literature because of limiting conditions. First, transaction costs are always positive and in this sense is yet another impossibility theorem (Blaug, 2000, p. 619). The most frequently raised arguments critical of R.H. Coase also include the lack of real opportunities in economic practice to estimate the level of damage caused by entities. Furthermore, it is relatively easy to identify the polluter, whereas it is very difficult (or even impossible) to identify all those affected.

Another issue directly related to the energy sector is that essentially all externalities that arise as a result of its operation involve public goods (ambient air, water), from the use of which no one can be excluded and solutions available to private entities cannot be implemented (Telega, 2010, p. 335).

Despite the criticisms, R.H. Coase's undoubted contribution to the emergence and inclusion of externalities in economic accounts is to draw attention to the fact that each such effect affects both parties. Therefore, it is not always justified to focus on one of them (usually the perpetrator). Furthermore, R.H. Coase signalled that there are opportunities to improve efficiency if conditions are created for the disputing parties to negotiate and contract. Only when the parties are unable to resolve the problem should the public authority use its sovereign powers.

3.3.3 METHODOLOGY FOR DETERMINATION OF EXTERNAL COSTS

The determination of external costs is based on Life Cycle Assessment (LCA). LCA is a recognised research method that identifies environmental risks. LCA is based on identifying and quantifying the amount of raw materials, energy and waste consumed and emissions of pollutants and then assessing the environmental impact of these elements. The LCA method recognises that every method of energy extraction and use has an impact on the environment in a certain way. These impacts can be characterised by different intensities and durations during the life cycle, covering all components, that is, natural resource use, production, packaging, transport, disposal and recycling. By considering a number of different impacts throughout the life cycle, it is possible to identify potential impacts that permeate from one life cycle stage to another or from one environmental problem to another. These are the main differences between LCA and other assessment methods, such as the carbon footprint (which only focuses on one environmental aspect) or methods that only focus on the direct emissions of products during use (Hauschild, 2018, pp. 59–66). The method also creates opportunities for consumers to make specific ecological choices with optimal environmental impact (Menten et al., 2013, pp. 108–134).

In order to compare the economic and social costs of generating energy from different sources and with different technologies, it is necessary to use analytic methods that allow costs to be normalised while introducing as many variables as possible into the study. Examples of methodologies commonly emphasised and applied worldwide include cost-benefit analysis (CBA) or the estimation of indicators using the LCOE or LACE method.

CBA allows all the costs and benefits of a project to be estimated from the point of view of the economy and society as a whole. It uses economic values that reflect the values that society would be willing to pay for a particular good or service. Economic analysis estimates all factors according to their utility value or an alternative cost to society. In CBA, the economic performance indicators of a project should be determined on the basis of the estimated cash flows, that is Maśloch (2011, pp. 309–314):

- **ENPV (economic net present value) of the project**—calculated by discounting all net economic flows associated with the project and summing them (formula 1). The ENPV indicates by how much the value of the expected net social benefits of the project (expressed in monetary units) exceeds the value of the inputs and future operating costs of the project.

Formula 1. Economic net present value

$$ENPV = \sum_{i=0}^{n} a_i S_t^E = \frac{S_0^E}{(1+r)^0} + \frac{S_1^E}{(1+r)^1} + \ldots + \frac{S_n^E}{(1+r)^n}$$

where
S^E—the balance of the economic flows of costs and benefits generated by the project in particular years of the adopted reference analysis period,
n—reference period (number of years),
a—economic discount factor,
r—the economic discount rate adopted.

- **The ERR (economic internal rate of return) of a project**—the discount rate at which the present value of the project's projected benefits and expenditure are equalised (at which the ENPV is zero) (formula 2).

Formula 2. Economic internal rate of return

$$ENPV = \sum_{t=0}^{n} \frac{S_t^E}{(1+ERR)^t} = 0$$

where
S^E—the balance of the economic flows of costs and benefits generated by the project in particular years of the adopted reference analysis period;
n—reference period (number of years).

- **B/C (benefit-cost ratio) of a project**—in terms of construction, it is similar to the Net Present Value (NPV). The B/C ratio is a measure of the relative return of a project's outlay, allowing one to determine how much profit (loss) the project will generate per unit of additional outlay. A project should be considered economically viable if the Benefit/Cost ratio takes a value greater than unity, meaning that the discounted benefits of the project are greater than the discounted expenditure (see formula 3).

Formula 3. Benefit-cost ratio of the project

$$B/C = \frac{\sum_{t=0}^{n} a_t B_t^E}{\sum_{t=0}^{n} a_t C_t^E} = \frac{\dfrac{B_0^E}{(1+r)^0} + \dfrac{B_1^E}{(1+r)^1} + \ldots + \dfrac{B_n^E}{(1+r)^n}}{\dfrac{C_0^E}{(1+r)^0} + \dfrac{C_1^E}{(1+r)^1} + \ldots + \dfrac{C_n^E}{(1+r)^n}}$$

Costs in the Energy Sector

where

B^E—the flow of economic benefits generated by the project in each year of the adopted reference period of the analysis,

C^E—the flow of economic costs generated by the project in each year of the adopted reference period of the analysis,

n—reference period (number of years),

a—economic discount factor,

r—the economic discount rate adopted.

In principle, any project with a negative ENPV, an ERR lower than the social discount rate or a B/C ratio less than unity should be rejected. A project with a negative ENPV consumes too many socially valuable resources for too a benefit that is too modest for the general public.

The distributed (levelised) cost of energy model (LCOE) is a microeconomic model. The use of the model allows a comparison of the cost of energy production from different sources (with unequal lifetimes, project sizes, different capital costs, risks, payback periods and capacities) and allows the perspective of the individual investor to be considered (Rhodes et al., 2017, pp. 491–499). Estimating the cost for each alternative energy technology considered comes down to expressing the LCOE (understood as the price for energy that would have to be charged over the lifetime of the plant to cover all costs—capital, operating and financial inputs) per kWh obtained (Kost et al., 2013, p. 36; Pawel, 2014, pp. 69–70). The LCOE can therefore be considered as the minimum cost at which energy should be sold (consumed) in order to break-even over the lifetime of the project. The general equation for determining the LCOE is presented in formula 4.

Formula 4. Distributed (levelised) energy production cost model

$$LCOE = \frac{NPV\ of\ Total\ Costs\ Over\ Lifetime}{NPV\ of\ Energy\ Produced\ Over\ Lifetime},$$

$$LCOE = \frac{\sum_{t=0}^{N} \frac{(I_t + M_t)}{(1+r)^t}}{\sum_{t=0}^{N} \frac{E_t}{(1+r)^t}},$$

where

LCOE—unit averaged lifetime cost of energy production (Euro/kWh),

It—investment expenditure in t-th year (Euro),

Mt—operational and financial expenditure in t-th year (Euro),

Et—energy production in t-th year (kWh),

r—discount rate (%).

Source: Own elaboration adapted from Bolinger et al. (2022), Heilmann and Houle (2011, p. 280) and Kost et al. (2013, p. 36).

However, the LCOE factor has many disadvantages, as it only considers initial and variable costs, as well as lifetime capital costs in the analysis. An LCOE-based analysis, however, does not take into account externalities, end-of-life costs or subsidies and tax credits. Despite this, it has become an indicator for cost comparison by policy makers, the society and businesses (Rhodes et al., 2017, pp. 491–499).

Due to its design, LCOE is most often used as a 'benchmarking' tool to assess the cost-effectiveness of different energy generation technologies. LCOE does not, in itself, reflect investment costs in a realistic way and should be used when comparing different energy projects with each other. Based on the construction of the LCOE indicator, as described in the previous chapter, we can conclude that the lower the value of the indicator, the more cost-effective and efficient the project.

Despite many disadvantages of LCOE not taking into account the full negative impact of conventional energy on socio-economic development, it is already clear from comparing LCOE for different technologies that key RES technologies are becoming competitive with new coal-based power plants and will become even more attractive in the coming years (see Figures 3.3 and 3.4).

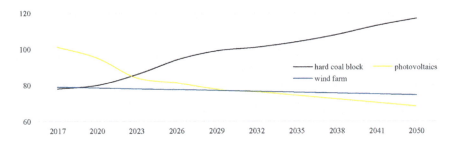

FIGURE 3.3 LCOE of RES variables and fossil-fuel power plant operating in load base from 2017 to 2050 (in Euro/MWh for r = 7%).

Source: Own elaboration adapted from J. Ecke et al. (2017, p. 29).

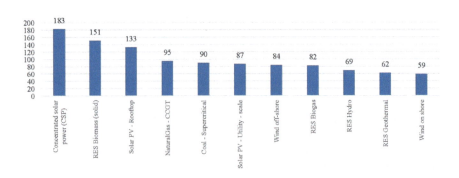

FIGURE 3.4 LCOE results for EU27 in 2018 (in Euro 2018/MWh).

Source: Own elaboration adapted from Cost of Energy (LCOE), Publications Office of the European Union, Final Report, 2020, p. 8.

A comparison of the costs of generating energy from renewable and conventional sources by LCOE shows that, by using RES, it is already possible to obtain energy at competitive costs. As presented in Figure 3.5, in the EU, LCOE for new projects show that most renewable energy sources have become cheaper than gas fired combined cycle gas turbines (CCGT) and supercritical coal power plants. In 2018, onshore wind LCOE were around €60/MWh, offshore wind around €85/MWh and utility-scale solar PV around €87/MWh (Final Report, 2020, p. 7). Given the investment pressures in RES development globally, it is possible to foresee a systematic decrease in the cost of installations generating renewable energy.

It is also important to bear in mind the shift among many countries—especially European ones—towards national and energy security. This is mainly due to Russia's use of energy resources as a tool for economic and political influence. For example, during the period preceding Russia's attack on Ukraine, the gas price crisis began. In the last quarter of 2021, prices for this commodity were in the range of 85–183 Euro/MWh (TTF), compared to 13–17 Euro/MWh (TTF) in the same quarter of the previous year. The crisis was mainly driven by two factors: a significant increase in demand for gas due to the economic rebound after the pandemic and a decrease in spot supplies from Russia. In its communication layer, Russia indicated the need to replenish its own gas stocks, assuring that long-term contracts were being fulfilled. This policy, portraying Russia as a reliable partner, was and still is a key element of the narrative and misinformation of societies and governments around the world. In an effort to counter Russia's aggression against Ukraine, the United States simultaneously made the information about preparations for an attack public. These factors caused great anxiety in the market and underpinned significant price volatility (Księżopolski et al., 2022, p. 200).

3.4 INSTRUMENTS FOR INTERNALISING COSTS

Internalisation of external costs can be done by:

a) environmental taxes and charges,
b) voluntary agreements,
c) ecological compensation,
d) fiscal reform,
e) deposit fees,
f) financial penalties,
g) direct regulation,
h) instruments based on market transactions,
i) subsidies.

3.4.1 ENVIRONMENTAL TAXES AND CHARGES

Taxes and charges are primarily defined as financial burdens imposed on producers or added to the price of a product. In principle, taxes and charges related to the internalisation of costs in the energy sector encourage entities to minimise the negative impacts of their activities, while they provide public authorities with the opportunity to take a strategic approach to planning the development and financing of energy activities.

In the literature, we find a detailed distinction between these concepts, and the authors of the definitions distinguish between these tributes mainly from the perspective of the entities that actually bear the costs of paying them, the funds they contribute to and the possibilities of expenditure (Slonimiec et al., 2015, p. 198). As a general rule, a tax is a gratuitous benefit, whereas a levy generates a claim to a specific benefit in favour of the payer on the part of public authorities and administrations.

In the European Union, taxes covering energy are part of a broader group of environmental taxes. According to the Regulation of the European Parliament and the Council,2 an environment-related tax is defined as 'a tax or charge based on a physical unit of something that has a proven, specific negative impact on the environment'. For EU Member States, the aforementioned Regulation imposes obligations of preparation of accounts for environment-related taxes, as well as reporting the data to the EC.

The European System of National and Regional Accounts (ESA 95), as an international methodological and accounting standard, characterises environmental taxes by physical units of impact with proven, unequivocally negative impacts on the environment. In this approach, the focus of the methodology is exclusively on environmental impacts resulting in the depletion of natural resources, environmental pollution and other forms of environmental deterioration, both locally and globally (Śleszyński, 2014, pp. 52–67). The criteria that qualify a liability as an environmental tax are presented in Table 3.4.

In accordance with the methodology adopted, four groups of environment-related taxes were distinguished, and the subject of taxation was used as a criterion for division (see Table 3.5)[3]:

1) energy taxes,
2) transport taxes,
3) pollution taxes,
4) taxes on the use of natural resources.

The number and types of taxes and forms of taxation vary from country to country. It is up to each country to select the appropriate tax instruments and their rates. The catalogue of environmental taxes adopted by a given country therefore depends

TABLE 3.4
Criteria for Qualifying a Liability as an Environmental Tax

Criterion for taxation	Criterion for environmental tax
At least one of the conditions laid down for taxes must be met the tax is not linked to the service or the tax payable is disproportionate to the service.	The environmental tax must have a documented detrimental environmental impact per physical unit of negative influence or its specifically identified proxy.

Source: Own elaboration adapted from Śleszyński (2014, p. 55).

TABLE 3.5
Environmental Tax Categories

Environmental taxes

Energy taxes:	Transport taxes:	Pollution taxes:	Taxes on the use of natural resources:
- Primarily taxes on energy carriers: fuel oil, natural gas, coal, electricity, etc.	- Taxes on the ownership and use of motor vehicles	- Taxes related to measured or predicted pollution discharged into the air and water (excluding CO_2)	- Taxes on natural resources exploited (e.g. mining tax, taxes on water intake and the use of forests, flora and fauna)
- Taxes on energy products used in transport such as petrol, diesel have been included in this group	- Also, taxes on other modes of transport, such as aircrafts and ships	- Also, taxes on solid waste management and noise, insofar as they meet the definition of taxes and the definition of environmental taxes	- Taxes on mineral oil, natural gas and taxes on land ownership are excluded
- This group also includes CO_2 emission taxes	- Taxes accompanying transport services (e.g. charters), insofar as they meet the definition of taxes and the definition of environmental taxes		

Source: Own elaboration adapted from Śleszyński (2014, p. 56).

on a number of factors, the most important of which are the level of socio-economic development, legal conditions, willingness to pay taxes, development priorities of the state or the level of environmental awareness of the population. The classification of a given tax into a relevant group is in principle dictated by its tax base. However, this is very often not clear-cut in economic practice. In such a situation, the 'majority rule' should be applied for a given tax category. Nevertheless, each tax included in the catalogue of environmental taxes, due to the interdependence of the individual categories, has a significant impact on the structure and functioning of the energy system in a given country.

The literature indicates that environmental taxes cannot provide public budgets with sustainable and significant revenues. The only exception includes energy taxes (especially on fossil fuels).

Environmental taxes can be divided according to different generic groups—for example, those related to energy (e.g. energy excise duty, customs duties or substitute charges for not purchasing the required amount of renewable energy), transport (e.g. vehicle excise duty, tax on means of transport, fuel surcharge), the environment (e.g. sewage charges, gas and dust emission charges, product charges) or natural resources (e.g. water intake charges, disposal charges).

When making comparisons of the level of tax rates and environmental tax revenues, particular care must be taken. On the one hand, for example, low tax revenues may signal relatively low environmental tax rates. On the other hand, they may be a consequence of high tax rates that have contributed to a change in consumption or activity to more environmentally friendly ones. Higher environmental tax revenues may also result from low tax rates that encourage non-residents to purchase taxed products abroad (as is the case of petrol or diesel) (Environmental Tax Statistics, 2022).

3.4.2 Voluntary Agreements

Contracts or voluntary agreements are types of arrangement between a central, regional or local authority and a sector (industry) in which a business or group of companies agrees to achieve a certain goal within an agreed time frame (Bauer and Fischer-Bogason, 2001, pp. 13–58; Iannuzzi, 2002, pp. 1–34; Nilsson, 1998, p. 26; Patronen et al., 2017, p. 46). Voluntary agreements work especially well in homogeneous sectors where there are not many small, dispersed producers or service providers. This is related, among other things, to the possibilities of mutual control in the implementation of the agreed objectives. Such arrangements also serve to minimise organisational, information or control costs (Landa, 1981, pp. 349–362).

In case of non-compliance, the public authority can impose regulation or other measures to prevent or enforce corrective action. In context of stimulating an increase in the use of renewable energy sources or improving energy efficiency, voluntary agreements should be considered as instruments equivalent to legal or financial regulation. A basic prerequisite for the successful achievement of the objectives of a voluntary agreement is the creation of appropriate incentive mechanisms by the public sector, encouraging economic operators to implement additional innovative or modernising measures. Economic practice shows that in most cases these are financial incentives and include direct financial or advisory assistance.

Voluntary agreements very often commit to measures that will result in a certain increase in the energy efficiency of production and the achievement of the indicated ecological effects. At the same time, the agreements are structured in such a way that a reduction in total energy consumption and a consequent increase in the energy efficiency of production are possible through the use of renewable energy sources (Nilsson, 1998, pp. A1–A15).

3.4.3 Ecological Compensation

Ecological compensation is an environmental management method that is defined in the literature as help to compensate for ecological damage caused during the development process. It aims to improve the environmental condition of the damaged area or to create (designate) a new area with similar ecological function and environmental quality (Yu and Xu, 2012, p. 892). Compensation can be paid directly to the victims or to public entities and comes from payments that are imposed under civil law on the perpetrator of the destruction. Such decisions force polluters to take measures to prevent or offset the amount of pollution. However, damage compensation requires basic conditions that are often considered impossible to meet (e.g. the problems of correctly assessing the damage, unambiguously identifying polluters and victims, and determining the cause-and-effect relationship between them, and finally creating a transparent and inexpensive procedure for the enforcement of compensation and payment).

3.4.4 Fiscal Reform

Fiscal reform involves reducing the fiscal burden on traditional factors (labour or capital) and increasing the fiscal burden on other resources, mainly energy. Fiscal

Costs in the Energy Sector 63

reform must therefore encompass the entire tax system in its scope and be directed towards environmental measures (Kudełko and Pękala, 2008, p. 17).

Environmental fiscal reforms are implemented especially in highly developed countries, where they attempt to solve the following problems (Śleszyński, 2001, p. 20):

- the inefficiencies of the existing tax systems, which do not stimulate labour productivity or encourage the efficient use of natural resources,
- the need for societies to bear external costs,
- the need to address regional and global environmental problems of reducing greenhouse gas emissions and natural resources depletion.

The fundamental challenge of fiscal reform is to create a system that should result in a double dividend. The first dividend to be achieved is environmental improvement. The second dividend, known as the socio-economic dividend, should be increased social efficiency (increased environmental awareness) and economic efficiency (use of revenues from public environmental tributes). The realisation of the double environmental dividend involves, among other things (Kudełko and Pękala, 2008, pp. 18–19):

- reducing external costs,
- creating technological change by introducing and developing environmentally friendly technologies,
- reducing the consumption subject to the resource tax (energy carrier).

The main problems of fiscal reform include the need to precisely identify the concept and scope of the reform, the environmental effects of implementing the reform or the impact on the competitiveness of entities and the economy. Reducing the fiscal burden of traditional factors in favour of increased taxation of energy consumption is expected to bring concrete, tangible benefits. First, it is expected to have a positive impact on the environment, and second, the tax-free income or profits saved by those adjusting to the fiscal requirements are expected to foster socio-economic development.

Green fiscal reform, both in terms of scope, implementation and effectiveness, is controversial. This controversy concerns environmental, financial, distributional, social or political consequences. The experiences to date of many countries implementing green fiscal reform do not provide clear-cut assessments—especially in terms of achieving a double dividend. The impact of the adopted solutions on employment levels and the level of competitiveness of economies also remains debatable. The implementation of an effective green fiscal policy also requires full social acceptance. At the same time, it should be noted that the increase in taxation of energy carriers is usually directly or indirectly passed on to consumers, which results in political resistance against the introduction of such solutions. In view of the fact that the introduction of any tax (especially on energy or fuels) is visible to consumers and directly leads to price increase, politicians are reluctant to decide on a green fiscal reform (Dresner et al., 2006, pp. 895–904).

3.4.5 DEPOSIT FEES

A deposit fee is a surcharge on the price of a product (to be kept by its purchaser) that is deemed to be particularly harmful to the environment. The deposit fee is refunded when the product is returned in a condition suitable for recycling, neutralisation or proper disposal. In economic practice, deposit schemes are primarily applied to consumer product packaging. They therefore include beverage packaging, glass bottles, plastic containers, metal cans and other durable packaging.

The characterisation of deposit charges from the point of view of the purpose can be considered from the ecological perspective of post-consumption behaviour or from the perspective of obtaining funds to finance the operating costs of deposit schemes. It should also be kept in mind that the price of the deposit-charged product must be substantial, creating an incentive for the user of the good to seek the return of the deposit.

Three basic types of deposit schemes are commonly used (Čekanavičius et al., 2003, p. 587):

1) company-initiated deposit schemes,
2) deposit schemes set up by public authorities,
3) security deposits (environmental pledges).

The use of deposit systems, although they do not directly relate to the energy sector, has a significant impact on the level of energy consumption in the economy. By returning a product that can be reused, it reduces the energy required to produce that product. In addition, it has a positive impact on environmental attitudes, forming the habit of saving and not wasting resources.

3.4.6 FINANCIAL PENALTIES

A financial penalty is a type of administrative sanction that is a financial incentive for law enforcement for non-compliance with certain norms. It boils down to issuing an order by a public administration body to order the payment of an amount specified in the act of law enforcement by an entity that has not complied with its administrative obligation (Staniszewska, 2015, p. 29).

One of the basic functions in the system of penalties is the preventive function that consists of depriving the entity on which penalties are imposed of a certain amount of money, which should induce that entity to take actions preventing it from paying these penalties in the future. As a rule, financial penalties are included in the category of non-deductible expenses, and the payment does not release entities from further obligations (resulting, for example, from the liability to make reparation or repair the damage caused).

3.4.7 DIRECT REGULATION

The use of direct regulation to internalise energy externalities is based on the administrative-legal regulation. Direct regulation can be divided into effect regulation (which

Costs in the Energy Sector 65

concerns the results of an activity (e.g. pollution)) and input regulation (which may concern, for example, bans on the use of certain resources, materials, techniques, technologies or compliance with quality standards). Direct regulation instruments include:

- introduction of environmental quality standards,
- setting emission limits,
- introduction of product standards,
- handling administrative and legal decisions,
- setting technological standards.

Direct regulation is popular and often used in practice. This is determined by the fact that public institutions in a position to introduce these regulations consider them to be effective. Furthermore, they are valued by interest groups because of the specific, lobbyable way in which they are introduced. An important factor encouraging representatives of public authorities to use these regulations is that they are readable by a wide social circle, among which they are recognised due to the belief that they are effective. They are also not commonly associated by consumers with the need to contribute to the costs of their introduction (Rumianowska, 2010, p. 132).

3.4.8 Instruments Based on Market Transactions

The main group of instruments based on market transactions are tradable pollution permits (emissions trading—allowance markets). In addition, other instruments may be used, such as, for example, pollution reduction credits received through joint implementation or certified emission reduction certificates.

Allowance markets as an instrument for internalising external costs are typically used in highly developed countries. They are characterised by both direct and indirect impact opportunities and are set up on markets where entities can purchase emission permits. The number of such permits is always limited, resulting in a situation where they are acquired via auctions (Van Egteron and Weber, 1996, pp. 161–173). It is widely noted in the literature that permit markets must be fully competitive and not controlled by any participant (Maedas, 2003, p. 294). Polluters are bound by limits, and when the amount of pollution is less than the allowed standard, an entity can sell the difference between actual and allowed emissions to another entity. The entity purchasing emission permits can increase emissions beyond the initial limit, which translates into the use of non-renewable resources.

Emissions trading can take various forms (Graczyk, 2009, pp. 104–105; Ranosz, 2008, pp. 86–87; Tietenberg, 1985, pp. 38–148):

- *Bubbles* mechanism—based on identifying and allocating maximum emission levels to companies so that they have the opportunity to undertake joint initiatives to effectively reduce emissions (can also operate within one company);
- *Offsets* mechanism—based on the investment made by companies, that increase emissions at their own plants, in facilities that reduce pollution at other companies;

- *Emission reduction credits*—involving the acquisition of reduction credits by companies with emissions below a certain threshold for resale to companies with emissions above that threshold;
- *Cap and trade*—involving setting of permissible limits of interference (pollution) in a given area, on the basis of which emission allowances are allocated. Companies operating in a given area must be entitled to a certain share of the interference with the environment. Allowances are granted free of charge by government authorities or are auctioned, and companies can use them in their current activity, sell them or keep them for future billing periods.

Emissions trading can operate at two levels (Jarno, 2016, pp. 128–129):

- *Upstream*—at this level, participation in the trade may involve a relatively small number of entities (e.g. producers and importers of fuels or energy raw materials) who will include in the prices of fuels and energy raw materials the incentives for end users to reduce emissions;
- *Downstream*—for this level, emissions trading is implemented at the point of emission and may involve a significant number of entities (e.g. power plants and cement factories).

The amount of emission allowances offered and available on the market is largely determined by their market price. One of the most developed emissions trading instruments has been introduced in the EU. In the case of the EU, allowances can be exchanged within the EU, with the result that the market price of an allowance is determined by supply and demand and defined by the marginal cost of reducing emissions. The main factors influencing the price of emission allowances in the long term is the amount of limits allocated by the EC under the National Allocation Plans. The legal basis for the European Emissions Trading Scheme (EU ETS) was initiated by Directive 2003/87/EC of the European Parliament and of the Council of 13 October 2003 establishing a scheme for greenhouse gas emission allowance trading within the Community, which began on 1 January 2005. The EU ETS is the world's first and largest international emissions trading scheme, covering three quarters of international emissions trading (EU Emissions Trading Scheme (EU ETS). An important element of the functioning of the EU ETS are the so-called 'trading periods (Directive 2003/87/EC of the European Parliament and of the Council of 13 October 2003):

1) in force between 2005 and 2007, had a test nature, which was reflected, among other things, in the fact that emission allowances from that period were not carried forward,
2) five years (2008–2012),
3) eight years (2013–2020),
4) to apply from 2021 to 2030.

The allocation of allowances under the EU ETS is implemented in two ways: through a free system (*grandfathering*), based on historical emissions, or an *auctioning* system, based on the sale of allowances (Eduarda et al., 2010, p. 48). The ETS is designed to reduce greenhouse gas emissions from energy used in industry. It gives businesses included in it a degree of flexibility, allowing them to choose between reducing emissions or acquiring emission allowances from other companies, depending on the market situation.

Directive 2003/87/EC has been modified on several occasions to extend the ETS to new industrial sectors or to strengthen incentives to reduce GHG emissions. This has led to a change in the nature of the scheme, which was originally designed to allocate GHG emissions cost-effectively among EU-wide emitters. It is now seen as a tool to promote the development of low-carbon investments. It should also be noted that from the very beginning of the introduction of the EU ETS, measures have been taken to increase the price of allowances in the market. These objectives are achieved, inter alia, by reducing emission limits and limiting the possibility of using substitute allowances (ERUs or CERs) (Dyduch, 2014, p. 68).

3.4.9 SUBSIDIES

Subsidies are a traditional form of funding, especially for activities with a positive impact on socio-economic development or the state of the environment. In their pure form, they are mainly made available to public sector units and social environmental organisations to finance the elimination of the effects of extraordinary environmental hazards, investments in nature conservation, etc. However, environmentally harmful subsidies (EHS) also exist in economies. These are mainly subsidies or tax breaks and exemptions that enable specific consumers, users or producers to supplement their income or reduce costs but have an overall negative impact on the wider environment.

Almost all countries apply direct and indirect subsidies (tax cuts and lower tariffs) for environmentally harmful activities. The use of EHSs consequently leads to higher levels of waste, emissions, resource extraction or to negative impacts on biodiversity. They can also perpetuate inefficient practices and hinder companies from investing in green technologies.

On a global level, the amount of EHS is estimated to be around €800 billion per year. Therefore, phasing out EHS or reducing them is an obvious opportunity to reduce external costs in the energy industry. According to the OECD, ending fossil fuel subsidies alone could reduce greenhouse gas emissions by 10% by 2050 (Wilts and O'Brien, 2019, p. 60).

NOTES

1 A.C. Pigou discussed the proposed tax in his 1920 book (Economics of Welfare), which is defined in the literature after his name as the Pigou tax.
2 Article 2 Regulation (EU) No 691/2011 of the European Parliament and the Council of 6 July 2011 on European environmental economic accounts (Text relevant for EEA) (OJ L 192, 22.7.2011, p. 1).
3 The basis of taxation is the physical unit (or a proxy for a physical unit) of 'something' that has a proven, specific and negative impact on the environment. The tax must at the same time be a mandatory and non-refundable financial burden. 'Something' that has an adverse effect on the environment is considered to be a phenomenon, action or object that affects the environment in a negative way.

4 Managing Transformation in the Area of Civic Energy

CONTENTS

4.1 Energy Transition Historically and Now 69
4.2 Effects and Challenges of the Current Energy Transformation—Strategic Directions of Development 74
4.3 Change Management and the Role of Leaders 81

4.1 ENERGY TRANSITION HISTORICALLY AND NOW

Energy transition means moving towards sustainable economies using renewable energy sources, energy conservation and increasing energy efficiency, in line with the principle of sustainable development. The ultimate goal of the transition is the complete replacement of non-renewable energy sources, mainly coal in the energy mix.

When analysing the problems of energy transition, it is important to emphasise that what drives the economic development of the world are great inventions, which in turn create breakthrough technologies for the use of energy. It was great inventions that triggered the first energy transition based primarily on fossil fuels.

The first energy transition was driven primarily by breakthrough inventions that hit the mainstream use, revolutionising the way people lived. A factor inextricably linked to the adaptation of new breakthrough technologies is the dynamic increase in energy demand. During the first energy transition, inextricably linked to the first and second industrial revolutions, solid natural fuels (wood and peat) were displaced by fossil fuels (coal and mineral oil). The beginning of the 20th century was thus a period of dominance of coal in the global energy mix, while the second half of the 20th century was a period of dominance of mineral oil.

The processes of the first energy transition enabled the industrialisation of the world, influencing the dynamics of economic development. Unfortunately, economic development and its dynamics have not been without impact on natural ecosystems in the form of overexploitation of natural resources, which in turn has contributed to environmental degradation and the accumulation of greenhouse gases in the atmosphere.

The negative effects of the first energy transition and the associated social costs have created a barrier to further global economic development.

Nowadays (since 1990), the so-called second energy transition has begun, which is also linked to breakthrough technologies and inventions such as the computer and the Internet. Their proliferation since the early 1990s has transformed these

DOI: 10.1201/9781003370352-5

inventions into all-encompassing digital solutions that are used universally, indicating that we are witnessing a dynamic adaptation of technologies and solutions that contribute to improving the living standards of societies. These include, for example, robotisation or electrification of transport, factors that have further fuelled the second energy transition, which clearly translates into an increased demand for clean energy.

The second energy transition therefore aims to break down the barrier resulting from the first transition and the change in the global economy and put it on a sustainable path.

It should be emphasised that within the first energy transition, it is possible to identify sub-periods that have significantly shaped the phenomena determining the second energy transition, as shown in Table 4.1.

The effect of the first and second energy transitions is an almost tenfold increase in energy demand, which in turn is shown in Figure 4.1.

The current energy transition is linked to the dynamic development of technology, the evolution of social preferences and (perhaps above all) a growing awareness associated with concern for the surrounding environment. The root causes of the dynamic increase in energy demand include several factors, which should be grouped into three main categories:

1. Technology development. The second energy transition is driven, like the first, by the development of breakthrough technologies, with this phenomenon occurring on an incomparably larger scale than the first transition. The main trend here will be the replacement of fossil fuel-powered appliances with renewable energy technologies.
2. Progressive concern for the natural environment and increased public awareness. In the second energy transition, the driving force for change is the environmental aspect and the increased need to take greater care of the

TABLE 4.1
First and Second Energy Transitions

The First Energy Transition (1850–1990)

Breakthrough inventions—the onset of the first transition

| 1860-combustion engine | 1880-light bulb | 1885-automobile |

Breakthrough inventions—developments in transport technology

| 1903-aircraft | 1930-jet engine |

Breakthrough inventions—economic growth, improving the standards of living

1973-mobile phone

Natural fuels in the form of wood and peat were displaced from the energy mix by fossil fuels—coal and oil. Energy was obtained by burning fossil fuels.

Second Energy Transition (1990–2022)

Breakthrough inventions—digitalisation, robotisation, electrification of transport, artificial intelligence.

| 1991-world wide web | 2007-iPhone |

Fossil fuels will be displaced by renewable energy sources (RES)—a process of transition from an era of burning fuels to an era of drawing energy directly from renewable sources and storing it efficiently.

Source: Own elaboration.

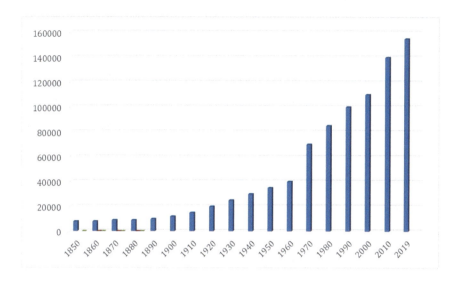

FIGURE 4.1 Growth in primary energy demand (TWh) from 1850 to 2019.

Source: Own elaboration.

planet. Which energy source will play a dominant role in the energy mix will be determined, as was the case until now, not only by the technological and economic aspect but also by an environmental criterion (at least in developed countries). The key role in the second energy transition will be played by those technologies and energy sources that can secure the ever-growing energy needs of societies, reduce the negative impact on the environment, and meet the criterion of economic viability of their use.

3. Social preferences—the third aspect of the second energy transition is based on changes in public awareness of environmental issues.

A consequence of the second energy transition will be the need for human activity to evolve towards complete independence from fossil fuels by converting naturally occurring energy into usable energy—electricity and heat.

A fundamental objective of the second energy transition is therefore to move away from fossil fuels. As Eurostat data shows, the share of renewable energy in EU countries varies—the data is shown in Figure 4.2.

For comparison, in addition to the EU countries, data for the leader in terms of RES use (Norway) and data for the United States are included (US data under RES includes renewable energy production and nuclear power). To get an even better understanding of what the second energy transition is and what challenges this process addresses, it is necessary to look at the projected degree of electrification of the mature EU economy by 2050. Such projection is shown in Figure 4.3.

The data presented in the figure once again proves that the demand for energy in modern, developed societies is an exponentially growing phenomenon, which in turn

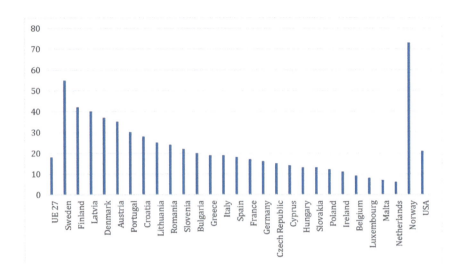

FIGURE 4.2 Share of energy from renewable sources in EU countries (as % of gross final energy consumption).

Source: Own elaboration adapted from (Environmental tax, 2022).

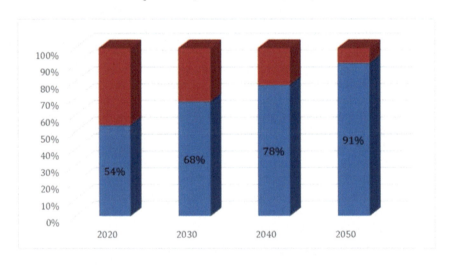

FIGURE 4.3 Degree of electrification of the mature EU economy by 2050.

Source: Own elaboration adapted from BloombergNEF (2022).

allows the claim that the use of fossil fuels for energy production will not meet global demand, and therefore the development of RES is at present the only strategic direction for the development of modern energy. Moreover, given the growing public awareness of the need to protect the environment, a discussion on civic energy should be initiated.

Civic energy, according to A. Dyląg, A. Kassenberg and W. Szymański, starts at the local—municipal—level. Every inhabitant of a municipality—a citizen of his or her small homeland—is an energy consumer. In turn, each municipality is obliged by law to provide energy to its inhabitants in places belonging to the municipality, for example. schools and offices (Dyląg et al., 2019, p. 9).

Civic energy is a system in which individuals, organisations, institutions and companies outside the energy sector take an active part in energy generation, transmission and management. The essence of civic energy is the local and small-scale production of electricity and heat from renewable sources and the use of energy efficiency solutions. Civic energy is also the participation of local communities in larger RES projects combined with the creation of local alternatives to a centralised and large company-dominated energy system (Dyląg et al., 2019, p. 11)

Civic energy is all about civic energy projects with tangible benefits for local governments, local communities, organisations and institutions, as well as for the local economy and the environment.

The most important of these include:

1) reduced energy demand—greater energy efficiency, resulting in significantly reduced heat and electricity bills;
2) increased energy security on a local scale—guarantee of the supply of the desired quality and eliminating interruptions in energy availability;
3) material and financial benefits for the local community;
4) social and economic development of the region—creation of sustainable green jobs, for example in the insulation of buildings or production and operation of RES installations;
5) lower emissions of CO_2 and other pollutants to air, soil and water—improved environmental quality and health benefits for residents;
6) strengthening social links through cooperation;
7) local energy independence.

A key determinant of civic energy is efficient energy management at the local level. Renewable energy sources, due to their small unit power and dispersed nature, are mainly connected to distribution networks. This is a great advantage, enabling the energy generated by RES sources to be used more efficiently locally, without the need to send it over long distances, that is, without putting additional strain on the network. The concepts being developed for energy clusters, energy communities and other collective forms of energy production and consumption will, in the future, allow for the creation of self-balancing areas which, by their very nature, will make more efficient use of existing network infrastructure. Managing the electricity grid at a local level in an energy cluster or energy cooperative also requires less investment through more efficient matching of generation sources to the demand. This also means that less effort is needed to increase the intelligence of local grids by investing in energy storage and appropriate computer-aided process management tools.

4.2 EFFECTS AND CHALLENGES OF THE CURRENT ENERGY TRANSFORMATION—STRATEGIC DIRECTIONS OF DEVELOPMENT

The strategic development directions for the energy transition and renewable energy are currently being set by China and the United States, which are leading the way with an influx of venture capital investment and numerous patents. It is worth emphasising at this point that the energy transition requires a technological breakthrough—almost half of the emissions reductions will have to come from technologies that are still underdeveloped.

Global renewable energy development is estimated to require 5.2 trillion USD of investment by 2050, according to current scenarios from the International Energy Agency—IEA (www.iea.org, 2022)

In order to achieve the goals of energy transition and decarbonisation, individual countries will have to go far beyond the mental and technological horizon we know today. Essential strategic projects will include:

1) exploring the potential of biofuels,
2) analysis of the feasibility of nuclear power,
3) developing methods for CO_2 capture,
4) further development of established renewable energy technologies such as solar and wind power,
5) decarbonisation of transport, food production and consumer goods production.

It should be emphasised that the aforementioned strategic transformations must be financed to such an extent that the price of low-carbon alternatives for consumers becomes equal to the standard market price. Such a change will induce the average citizen to act towards sustainability (www.ey.com, 2022).

Under the conditions of contemporary globalisation and global warming, moving away from fossil fuels is the immediate necessity. The aforementioned problem has become even more topical following the outbreak of armed conflict in Ukraine—we are currently dealing with an unprecedented energy crisis arising from the actions of the Russian Federation in restricting supplies of these raw materials to the world market. Such a situation shows clearly how important it is for the economies of developed countries to become independent of fossil fuels.

It is worth noting at this point that the functions of leaders of the change towards renewable energy are only served by a few countries, which thus become the biggest beneficiaries, due to the fact that they attract most of the money stream for renewable energy development. They are the ones who can accelerate the energy transition globally. Thus, the United States is not only an opinion-forming leader thanks to its strong academia and numerous publications on renewable energy, but it also has a significant advantage over other economies in terms of corporate investment in renewable energy. China, in turn, leads in terms of installed renewable energy capacity and material supply for sustainable transport. Over the past 10 years, China has spent more than 58 billion USD on electric vehicle subsidies, giving it the capacity to produce half of the world's electric vehicles and attracted a significant amount of venture capital funding for the sector.

China's position, alongside countries such as the UK, Germany and France, is strengthened by a range of government support instruments targeting renewable energy development. Thanks to, among other things, feed-in tariffs to promote the development of distributed solar power installations and subsidies to support the development of offshore wind power, China's renewable energy capacity has increased to 42% of the country's total generation capacity. The value of investment by public entities in China in green ventures averaged 320 billion USD annually in 2018. Such instruments are drive the development of a country that already has 32% of global renewable energy capacity.

The investment gap between the developed and developing world in renewable energy is huge. G20 countries invest almost 50 times more in renewable energy than the rest of the world in early-stage growth ventures. Unfortunately, these countries are still heavily dependent on fossil fuels, which consequently hampers the development of renewable energy. As discussed so far, the key to achieving the energy transition goals will involve a technological leap. The ongoing second energy transition requires and will require effective cooperation at both global and local levels.

Among the strategies related to the implementation of the new energy deal, the following actions should be identified:

1. **Accelerating the climate action**. National governments should support processes to transition from fossil fuel-based energy to 'green energy' as quickly as possible. Such action should be intensified through a system of incentives to increase the production of proven green technologies, the transformation of companies towards net-zero emissions or support for business incubators. The financial sector should play an important role in the green transition by promoting a so-called green financing. Businesses, in turn, should increasingly support change by meeting net-zero emissions targets, developing innovative technologies and disseminating products and services that provide affordable alternatives to existing offers.

2. **Responding to the new global climate challenges of the 21st century.** Climate change is increasingly visible. A fundamental challenge of the 21st-century societies is to do everything possible to reduce the rate of climate warming. Despite initiatives in this area, it must be recognised that the pace of change is not sufficient. In order to accelerate the transformation, there is an urgent need to increase the involvement of developed economies, governments, financial institutions and citizens, and to dramatically increase investment in research and development, business investment, the development of an industrial infrastructure supported by public policy and consumer involvement, that is, the building of a so-called civic energy.

Despite the carefully prioritised efforts to move away from fossil fuels and drastically reduce CO_2 emissions, phenomena are emerging in the global world of the 21st century that go completely against the climate threats. Such phenomenon is the ongoing since 24.02.2022 war in Ukraine, which, in addition to annihilation of the Ukrainian population and the war damage, represents a fundamental burden on the climate—during warfare, no attention is paid even to the level of CO_2 emissions.

Another indication of the need to implement the demands of the second energy transition as soon as possible is the energy crisis resulting from lack of fossil fuel imports from Russia. The aforementioned energy crisis becomes a proof that the energy transition is an inevitable phenomenon, while the use of RES is basically the only way to obtain green energy and become completely independent from Russian supplies.

An important approach in the area of strategic management concerning the development of RES and hydrogen technologies is the identification, by experts, of ten areas of intervention in the energy transition aimed at green energy. In addition, the verified areas take into account the social aspects of the transition, emphasising the role of local governments in the process (https://energiapress.pl, 2022). The energy transition is a desired process not only because of catastrophic climate change, but it is also essential in a form of technological revolution based on building an alternative green economy with green transport and industry. The future is electrification of all sectors of the economy based on zero-carbon sources. High fossil fuel prices and the risk of supply disruption will accelerate the energy transition and the development of new technologies—in transport, heating, industry. Energy security will have a new dimension—it will be based on ensuring a stable and reliable energy supply based on local, renewable sources.

The aforementioned ten areas of intervention in the energy transition in Poland include (https://zielonerozwiazania.pl, 2022):

1. A GOOD PLAN FOR ENERGY TRANSITION IN POLAND: Transition must be conducted wisely. It must occur through evolution, not revolution. The change should start with the energy sector and then extend to the whole economy, analysing the benefits and also looking for potential risks.
2. RES AND NUCLEAR ENERGY DEVELOPMENT: For many years, the development of renewable energy sources has been motivated in Poland by the need to implement EU energy and climate policy. Nuclear power is, next to wind power, the resource with the lowest lifetime CO2 emissions. Both nuclear and RES should complement each other to jointly create a zero-carbon economy of the future.
3. NETWORK MODERNISATION: The function of energy grids is to provide a service and enable the execution of the agreement between a consumer and a producer—a very important statement that should make it clear that all activities carried out in the area should be directed towards performing this function.
4. ENERGY INDEPENDENCE FOR LOCAL AUTHORITIES: One of the biggest challenges of the transition is the change from a system based on large generation sources to a dispersed structure in which local energy sources will play a leading role in securing the energy needs of energy consumers.
5. ELECTRIFICATION OF THE ECONOMY: One of the main ways to achieve climate neutrality is to electrify the economy based on renewable energy. In the short term, the definition of what we consider expensive will also change, for example, storage, green hydrogen will prove to be widely affordable compared to expensive coal and gas. Renewables, especially wind and solar with zero fuel cost—will be the most desirable form of investment—both by the public and financial institutions.

Managing Transformation in the Area of Civic Energy

6. THE DEVELOPMENT OF HYDROGEN TECHNOLOGY: In these difficult times, the question of how Europe can quickly become independent of raw materials from Russia is increasingly being asked. The answer to this question has been known for a long time and it is 'hydrogen', which as a gas can successfully replace fossil fuels in the transport, energy or chemical industries.

7. RESIGNATION FROM COAL, CARE FOR THE CLIMATE AND THE ENVIRONMENT: Coal is the main energy resource in Poland, combustion of which generates 70% of electricity and over 90% of heat. Decarbonisation means not only the need for new generation capacity but also the elimination of economic monocultures in places where coal is currently extracted and burned.

8. SUSTAINABLE INVESTMENT AND NEW TECHNOLOGY DEVELOPMENT: The green revolution is well underway in Europe. Energy transformation, zero-carbon, closed-loop economy are real goals for which huge funds need to be allocated.

9. BUILDING A GREEN ECONOMY: Building a green economy is not just about reducing emissions and energy intensity. One of the intrinsic elements of the changing landscape of Polish energy sector will be a long-term transformation of job characteristics, the technologies and services used, as well as building of competence in research and development.

10. EDUCATION: The energy transition, like any great transformation, must receive public acceptance in order to be successful. No great social and economic reform is fully possible without understanding and the need to work towards changes. For this to happen, there is a need for widespread education through activities in schools, universities and also in public spaces.

The principles of the decalogue indicated earlier are, of course, universal and can be applied in other countries undergoing a second energy transition. It should be emphasised that in the contemporary realities of the global economy, the energy crisis and the uncertainty in the financial markets related to the war in Ukraine, the implementation of the decalogue will be an extremely complicated task. Moreover, at the moment, we do not have the technology to build 100% energy independence based solely on renewable sources.

When analysing energy transition processes from a strategic perspective, it is worth using one of the portfolio methods for conducting a strategic analysis, which includes the SWOT analysis. In connection with the role of civic energy in the future energy mix, which has been emphasised several times, the SWOT analysis was oriented towards presenting the strengths and opportunities associated with support for the development of civic energy. In turn, the weaknesses and threats present barriers to the development and implementation of the civic energy concept. The SWOT analysis was carried out with a breakdown into three segments: prosumer, community and local government.

Table 4.2 presents a SWOT analysis in relation to the prosumer segment.

Table 4.3 contains the SWOT analysis carried out for the community segment.

In turn, Table 4.4 proposes a SWOT analysis for the local government segment

The proposal to carry out a SWOT analysis divided into prosumer, community and local government segments provides an interesting concept for a strategic analysis

TABLE 4.2
SWOT Analysis in Relation to the Prosumer Segment
The Prosumer Segment

Strengths

- The ability to balance consumption and production of electricity in the network;
- The stability of the support system in the so-called 'discount' system (in contrast to the unstable certificate system that worked before and the FiT system, which also requires constant changes in the level of support);
- Facilities for the technical installation of RES sources up to 40 kW (exemptions from building permits and other);
- Widespread public interest in installing micro-installations, stable public support for RES development and a willingness to install RES in one's own home at >80% of the country's population;
- Educated technical staff to support the prosumer movement (but not sufficient).

Weaknesses

- Lack of knowledge of the procedures for connecting the installation, lack of a single place where they can easily be learned (which is why prosumers use installers or energy advisors in this regard);
- No possibility to sell surplus electricity to the grid—this electricity is confiscated by the grid operator;
- Electricity grid problems affecting the prosumer, e.g., lack of devices to support the prosumer's own power production in the grid, mainly bi-directional meters (to be installed only when the RES source is connected—this prolongs the procedures);
- Lack of solutions/legislation to enable the use of surplus heat from prosumer installations in local heat networks;
- Lack of possibilities/regulations to exchange electricity with other users of the electricity grid;
- Lack of or very low support for prosumers from municipal communities (local authorities), as they have few capacities in this area, despite planning obligations, and very limited staff (e.g. frequent lack of a municipal energy officer);

Opportunities

- Tightening climate policy globally and in the EU will create increasing incentives for RES development including prosumer energy;
- Growing political and non-political support from the EU regarding the development of energy democracy;
- An expected high share of funds for the development of a low-carbon economy in the new EU budget—25%;
- Access to funds to subsidise installations by at least 50% (reducing payback time to 5–10 years);
- Rising electricity prices in electricity grids (this increases the cost-effectiveness of installing own power supply);
- A growing lack of acceptance of dirty air motivating the purchase of RES heat installations;
- The increasing interest of local municipalities (politicians) in supporting the development of micro-installations, as a result of public interest that can translate into political support.

Risks

- A reduction in the level of subsidies for RES installations in future years, due to their increasing profitability;
- Stabilising electricity prices at a low level through a deliberate state policy as a shield to electricity consumers;
- The rapid growth of the energy industry increases the risk of fraud;
- Politically limited opportunities for the development of prosumer installations due to the status quo in terms of power generation from large power plants, including coal and gas (in prospect perhaps nuclear).

Source: Own elaboration adapted from Kassenberg and Szymański (2019, p. 97).

TABLE 4.3
SWOT Analysis Carried Out for the Community Segment

Community Segment

Strengths

- Attempt to regulate the sphere of energy clusters and cooperatives in law—creating opportunities for their operation;
- Moderate interest in ways to support RES in other countries, e.g., Germany, in the form of, e.g., cooperatives;
- Stable public support for RES development and willingness to install RES in one's own home at >80% of the country's population;
- Ownership of local energy and heat networks, e.g., within housing associations;
- A growing crowdfunding movement that can support community RES initiatives.

Weaknesses

- Low cooperation and investment activity at the level of housing communities due to the need for consensus within the residents' group;
- Lack of knowledge of the procedures for connecting the installation, lack of places where this can be easily learned (which is why prosumers use installers);
- Unclear status of the use of electricity, e.g., from an installation in a housing association, for individual dwellings;
- Unclear status of the possibility of creating closed/separate energy networks in the community formula;
- Lack of possibilities/regulations for exchanging electricity with other users of the electricity grid without its intermediation;
- Lack of technology offers targeted at community solutions.

Opportunities

- Growing support from the EU for the development of energy democracy;
- Planned development of support systems for collective production (cooperatives, clusters);
- The possibility of creating hybrid systems involving, for example, photovoltaics, a heat pump and an electric car as storage.

Risks

- A reduction in the level of subsidies for RES installations in future years, due to their increasing profitability;
- Stabilising electricity prices at a low level through a deliberate state policy as a shield to electricity consumers;
- Politically limited opportunities for the development of prosumer installations due to the status quo in terms of power generation from large power plants, including coal and gas (in prospect perhaps nuclear);
- Existence on the market of an alternative to the community/cooperative solution, in the form of a so-called: dispersed shareholding of companies investing in large-scale RES.

Source: Own elaboration adapted from Kassenberg and Szymański (2019, p. 99).

concerning the introduction of RES solutions. It is worth emphasising that a reliably conducted SWOT analysis and adherence to the provisions of the analysis will allow, on the one hand, for the rational use of opportunities and the exposure of the strengths of the enterprise (organisation) conducting the energy transformation, and on the other hand, for the avoidance of threats and the lack of exposure of weaknesses.

TABLE 4.4
SWOT Analysis for the Local Government Segment
The Self-Government Segment

Strengths

- Energy planning at municipality level is mandatory and can lead to the stimulation of local prosumer initiatives;
- There are good examples of municipalities that are active in the energy market and have their own local energy policy;
- Ownership or co-ownership by municipalities of energy and heat or electricity networks (e.g. tram transport networks);
- The possibility of using the municipality's own resources, municipal enterprises and aid funds for energy management in the municipality;
- Experience of municipalities in implementing subsidy programmes aimed at citizens in the field of RES and in the implementation of RES investments;
- The possibility for the municipality to gain financial benefits from commercial RES investments in the municipality (e.g. large installations).

Weaknesses

- The development of infrastructural investments in local energy networks is practically outside the competence of the municipality (the exception is the co-ownership of district heating networks);
- Lack of up-to-date information on RES installations and energy efficiency at municipality level (reporting is done directly to the central level);
- Municipalities have few energy capacities, despite planning obligations, and very limited staff (e.g. frequent lack of a municipal energy officer);
- Particularly in large cities, the operation of private thermal power companies, on which municipalities have limited influence.

Opportunities

- Growing political and non-EU support for the development of energy democracy;
- The anticipated high share of funding for the development of a low-carbon economy in the new EU budget (e.g. 25%);
- Increasing electricity prices in the Polish grid (this makes it more profitable to install own power supply);
- A growing lack of acceptance of dirty air motivating municipalities to set up heat source replacement programmes;

Risks

- Continuation of the monopolistic, pro-coal policy of the Polish government supported by circles gathered around large-scale 'dirty' energy with considerable political power;
- Centralisation of energy policy measures aimed at limiting the development competences of local authorities (e.g. the sphere of spatial planning);
- Weakening of the EU's pro-climate policy, whether as a result of poor progress in international negotiations or resistance within the EU;
- Lack of respect of international agreements by the signatories, i.e. the Paris Agreement;
- Politically limited opportunities for the development of prosumer installations due to the status quo in terms of power generation from large power plants, including coal and gas (in prospect perhaps nuclear).

Source: Own elaboration adapted from Kassenberg and Szymański (2019, p. 100).

Managing Transformation in the Area of Civic Energy

4.3 CHANGE MANAGEMENT AND THE ROLE OF LEADERS

The processes of designing and implementing the energy transition are characterised by the fact that societies are particularly sensitive to the effects of these measures, which are very often mutually exclusive. An example of such exclusion is the abandonment of fossil fuels—on the one hand, this is the immediate necessity, which has been mentioned several times, but on the other hand, the effect of such measures will be to increase unemployment in the mining sector. Thus, energy transition processes involve constant decision-making, whether international (e.g. within the EU), national, regional (civic energy) or even within companies. Decision-making is therefore an inherent part of management action in the energy sector, which involves the deliberate and conscious choice of one of the possible courses of action (Zdyb, 1993, p. 75). Etymologically, the term *decision* comes from the Latin language and means a decision, an adjudication and a resolution. A decision is an act of free choice (Bartkowiak, 2009, p. 160). Analysing the approach of individual authors to defining the concept of *decision-making*, it should be noted that it is a process (analytical, verification, monitoring) aimed at selecting (from among several options) the optimal solution, which, in the opinion of the decision-maker, guarantees obtaining the intended results (Bartkowiak, 2009, p. 160).

It is worth noting that the different terms emphasise conscious and rational choice. This approach means that since the choice is conscious, it cannot be random. In order for there to be a choice, the decision-maker must have at least two options for solving the problem. But in addition to the alternatives and the possibility of an informed and rational choice, they must express the will to make a decision. Hence, decision-making should be understood as the course of the decision-maker's thought processes leading to the selection of one out of two or more options. Hence, decision-making has a certain specificity resulting from a sequence of activities shaped in such a way that they require consideration of the temporal and spatial structure in the implementation of the tasks to be achieved and the selection of appropriate courses of action.

Among the decision-making problems, a distinction is made between:

1) real and apparent problems (answering the questions: what and how it relates?),
2) operational and strategic issues (answering the questions: what to achieve, how to achieve it and with whose help?).

Management decision-making is a process consisting of a series of consecutive phases including, but not limited to, the following (Pietraszewski, 2004, p. 153):

1. **Identification of the purpose of the decision and the problem to be solved**. The decision problem arises from the current situation that needs to be solved in the best way for the organisation's objectives. In defining the problem, it is important to note the complexity of the problem and

the conditions, knowledge of which can be useful to solve the problem effectively.

2. **Exploration of alternative solutions**. Based on the information gathered, several options for solutions can be identified that may contribute to achieving the objective. It is important to anticipate the economic and social impact of each of the proposed solutions.

3. **Comparison and analysis of the alternatives.** Various solution options indicate different methods and means of action to achieve the objectives. In this phase, they are comprehensively analysed and compared with each other in order to make the best possible choice.

4. **The act of choosing**. At this stage, the manager makes a choice from a set of alternatives. The decision made may coincide fully or with some modifications with one of the proposed solution alternatives. It may also be different from the prepared alternatives if the decision-maker considers it necessary.

5. **Direction of implementation of the decision**. This phase is where the agreed action is initiated and implemented.

6. **Supervision and control.** This is the stage to ensure that the result is in line with the objectives adopted. During supervision, it is possible to partially modify the decision taken, if necessary.

According to the division criteria used, several types of decisions are distinguished:

1) according to the criterion of time of decision:

- concerning topical issues (operational),
- concerning developmental problems (strategic).

2) according to the degree to which the decision-maker is informed, a distinction is made between decisions:

- undertaken under conditions of certainty,
- undertaken under conditions of risk,
- undertaken under conditions of uncertainty,
- undertaken under conditions of incomplete information,
- undertaken in a conflictual environment.

The phases of decision-making are closely linked and harmonised.

The analysis of the decision-making process shows that the basic condition for efficient decision-making is obtaining complete, rapid and reliable information, which is the essence of the management process.

Information, taken as a signal that increases our knowledge of reality, is transmitted in the process of communication. Information can be divided according to different criteria. The basic classification is the division into primary and derived (secondary) information. Primary (analytical) information undergoes a so-called reduction (processing) through selection and aggregation and is then transformed into derived information. An important issue for the efficiency of the decision-making process is the reliability of information and, in particular, protection against so-called information noise.

Another key issue concerning the process of change and energy transition is precisely change management. In this view, change can be defined as a conscious and deliberate action, the essence of which is the transition from an existing state to a new state that meets the objectives and is able to fulfil the new tasks.

Rapid technological change, globalisation of markets, and competition or climate change have led to a situation where, paradoxically, the continuity of change is basically the only factor that does not change. In the aspect of revolutionary changes, and energy transition processes are among such, it is important to emphasise the existing, natural resistance of society to change (Stoner et al., 2001). Even the best ideas, absolutely necessary solutions, are met with rejection simply because they are new. Overlooking this fact is the most common reason for the difficulties, delays or even inability to actually achieve the benefits that a change can bring.

Several ways of dealing with the indicated resistance to change are currently proposed in management theory:

1) Informing the public of the need for change before it is implemented. The more diverse the forms of this communication, the more effective they are (e.g. financial data, expert statements, scientific opinions).
2) Paying particular attention to the motivation for change of groups or individuals who can potentially have a major impact on society. Different social groups or even interest groups should be kept in mind.
3) Presenting as accurately as possible the system of change, its consequences, the new ways of doing things, so that the 'new' is as little unknown as possible.
4) Consistent, decisive action on the part of change-makers.
5) Planning change as a process rather than a one-off event.

Public resistance to change is mainly due to underreporting or a feeling that decision-makers are manipulating information and revealing only part of the information regarding the consequences of making changes or the consequences of not making changes. In the case of fundamental and strategic changes, one of which is the energy transition, people crucial to the success of the whole process are the leaders. It is them who, using available resources and by building public campaigns, must convince people of the need for change by discussing transformation processes in the language of future benefits. The biggest disadvantage of the energy transition management activities proposed earlier is that they are time-consuming. However, it should be emphasised that in the case of ground-breaking activities, the basic criteria are efficiency in implementation, changes in public awareness, future economic efficiency, energy security and the prospect of clean energy.

5 Conditions for the Development of Renewable and Civic Energy

CONTENTS

5.1 Definition, Division and Characteristics of Renewable Energy 85
    5.1.1 Re a) Solar Energy 86
    5.1.2 Ad b) Wind Energy 87
    5.1.3 Ad c) Geothermal Energy 87
    5.1.4 Re d) Hydropower Energy 87
    5.1.5 Ad e) Biomass 88
5.2 The Importance of Renewable Energy for the Economy 88
5.3 The Impact of Renewable Energy on the Natural Environment
    and Human Health 93
5.4 Influence of Social Awareness on the Possibilities and Directions
    of Development and Use of Renewable Energy Sources 95
5.5 Directions and Trends in the Development of Renewable Energy—the
    Key Role of Civic Energy 97

5.1 DEFINITION, DIVISION AND CHARACTERISTICS OF RENEWABLE ENERGY

There is no definition in the literature or legislation that uniquely characterises renewable energy (see Table 5.1). Definitions are highly dependent on particular countries energy policies and the scale and quality of the academic debate. The most significant differences (debatable issues) relate to the classification of hydroelectric power plants (depending on capacity) and biomass derived from waste as renewable energy (Jordan-Korte, 2011, p. 11).

Analysing the aforementioned definitions, renewable energy includes:

a) solar energy,
b) wind energy,
c) geothermal energy,
d) hydropower energy,
e) energy generated from biomass.

DOI: 10.1201/9781003370352-6

TABLE 5.1
Definitions of Renewable Energy

Author	Definition—description
Lund	Energy that is produces from natural resources (sun, wind, rain, waves, tides and geothermal).
Dictionary of Energy, Cleveland, Morris (Eds.).	Energy source derived directly from solar energy (e.g. PV), indirectly from the sun (e.g. wind, biomass) or from other natural energy flows (e.g. geothermal).
The Australian Renewable Energy Agency (ARENA)	Energy which can be sourced from natural resources (can be constantly replenished).
European Parliament	Energy from renewable non-fossil sources (wind, solar geothermal, hydrothermal, hydropower, energy from biomass, landfill gas, sewage treatment plant gas and biogas).
J. Twidell, T. Weir	Energy that comes from natural and reproducible energy flows (occurring in the local and regional environment).
National Geographic	Energy obtained from sources that are inexhaustible and replenish themselves naturally.
International Energy Agency (IEA)	Energy produced by processing natural processes (e.g. solar, wind, biomass, geothermal energy, hydropower and biofuels).

Source: Own elaboration adapted from: Lund, 2004, p. 8; Twidell and Weir, 2015, p. 3; Renewable energy sources and climate change mitigation. Summary for policymakers and technical summary. Special report of the intergovernmental panel on climate change. Intergovernmental Panel on Climate Change, Potsdam, 2012, p. 166; International Energy Agency, 2009, p. 17; Cleveland and Morris, 2005, pp. 371–372; Article 2, Directive 2009/28/EC of the European Parliament and of the Council of 23 April 2009 on the promotion of the use; www.eia.gov/energyexplained/index.cfm?page=renewable_home, 2022; www.nationalgeographic.org/encyclopedia/non-renewable-energy/, 2022.

5.1.1 Re a) Solar Energy

Solar energy is one of the inexhaustible energy sources available in every part of the world. Solar energy is the energy from the sun's rays converted into thermal or electrical energy. Next to wind, solar energy is one of the oldest energy sources used by humans. In modern times, the past 3–4 decades have seen a period of increasing interest in ways of harnessing and utilising the potential of solar energy. This is related, among other things, to the increasing technical possibilities of converting solar energy into electricity (Foster and Ghassemi, 2010, pp. 7–40; Jones and Bouamane, 2012, pp. 1–86).

The amount of solar energy reaching the Earth is enormous, estimated at around 2.9 million EJ/year. This therefore means that it exceeds the entire humanity's total current energy requirements more than 5,000 times. Although the technically exploitable potential is much smaller and is already only about 1,500–50,000 EJ/year, it is still many times greater than the current global energy demand (Matuszczyk et al., 2015, pp. 183–186). The potential for sourcing solar energy varies regionally (World Energy Resources Solar, 2016, pp. 7–8). Solar energy is used in several ways

Conditions for the Development of Renewable and Civic Energy

in modern economies to produce electricity or heat (through photovoltaic cells to produce electricity, solar collectors to produce heat or thermal solar power plants).

5.1.2 AD B) WIND ENERGY

Wind energy is defined as the kinetic energy of the wind used to produce electricity in wind turbines (Energy from renewable sources in 2008, GUS, 2009, p. 10).

In order to harvest energy from the wind, wind power plants are used, which consist of devices that produce electricity using wind turbines. Wind energy can be developed as part of onshore wind energy (both as large-scale wind energy—and small-scale (dispersed) wind energy) and offshore wind energy (wind farms located in open sea).

A significant disadvantage of wind power is its practically total dependence on weather conditions, the fluctuations of which do not allow wind turbines to contribute to electricity generation to the extent to completely eliminate its deficit.

5.1.3 AD C) GEOTHERMAL ENERGY

Geothermal energy is heat extracted from within the earth in the form of hot water or steam. Today, use is made primarily of the heat stored in water deposits (low-enthalpy deposits used directly to produce thermal energy, characterised by temperatures below 150°C) and geothermal steam (high-enthalpy deposits used to produce electricity, characterised by temperatures above 150°C), which carry the energy to the surface. Increasingly, heat is also being recovered by means of heat pumps from shallow parts of the earth's crust, where heat with temperatures of up to several degrees Celsius is extracted. In the future, *hot dry rocks* that do not contain steam or geothermal water will also become prospective sources of heat (DiPippo, 2012, pp. 443–478; Stober and Bucher, 2013, pp. 165–202).

Geothermal energy may play a special role in the future, especially in Europe. This is due, on the one hand, to the significant potential of technically exploitable geothermal resources and, on the other hand, to the technological sophistication and capital equipment of European countries. In addition to increasing its share in the production of electricity, heat or cooling, geothermal energy can become an important factor influencing energy security, especially in terms of alleviating problems related to external natural gas supplies from the Russian Federation (Renewables for heating and cooling and EU security of supply: save over 20 billion euro annually in reduced fossil fuel imports, 2014, p. 5).

5.1.4 RE D) HYDROPOWER ENERGY

Hydropower is one of the oldest energy resources, successfully used by humans for millennia and remains competitive with other energy sources for many reasons.

If large hydropower plants are controversial (especially environmentally), small hydropower plants can be developed in both environmentally and socially sustainable manner. Globally, no uniform definition and characterisation of small hydropower plants has yet been developed. Depending on the country, this group includes power plants with installed capacities of up to about 1.5 MW and even up to 25 MW

(Paish, 2002, p. 538). In the case of Europe, there is also no uniform definition and there are many divisions, where installed capacity is considered as the upper limit: 1.5 MW (e.g. Luxembourg or Sweden), 5.0 MW (e.g. Poland, Austria, Germany) or 10 MW (e.g. Belgium, Spain) (Wiatkowski and Rosik-Dulewska, 2012, p. 316). In addition, the development of technologies to extract energy from the seas and oceans also offers great opportunities in the future (Melikoglu, 2018, pp. 563–573).

5.1.5 AD E) BIOMASS

Biomass has been used for energy purposes for thousands of years and has a significant impact on the energy system as well as on agriculture, the environment or other areas of socio-economic life (Andreae, 1991, pp. 3–21). According to the European Union, biomass is biodegradable fraction of products, waste and residues from agriculture, forestry and industrial and municipal waste (D Directive 2001/77/EC of the European Parliament and of the Council of 27 September 2001 on the promotion of electricity produced from renewable energy sources in the internal electricity market, Official Journal L 283, 2001). Biomass therefore mainly consists of residues and waste. However, some forms of biomass are a goal rather than a side-effect of production. For example, crops with a high annual growth rate and low soil and water requirements are grown specifically for biomass. The main sources of biomass used for energy purposes include varieties of forest trees and energy crops.

As with any agricultural activity, biomass production can have negative impacts on socio-economic development and the environment (Andreae, 1991, pp. 3–21; Mohtasham, 2015, p. 1289–1297). From the analysis of the literature on the subject and the available research results, it is clear that one of the key problems of biomass energy production is its huge land requirements (Popp et al., 2014, pp. 571–575), which results in many conflicts. These concern land used for production for food as well as other activities (paper industry, infrastructure development, environmental protection, etc.) (Staliński, 2016, pp. 99–100). Waste from agriculture, agri-food industry or municipal waste should be used for energy purposes in the first place. In addition, in all rural areas there are wastelands, areas of land that can be used to produce biomass for energy purposes. What remains unsolvable, however, is the problem, as with any land use, of energy crops competing with other renewable energy installations and other socio-economic objectives.

An analysis of the use of land for agricultural purposes clearly shows that agricultural yields are steadily improving. In combination with the negative demographic situation in many highly developed countries, this is leading to a situation where there is less and less need for agricultural land used for food. In such a situation, production for biomass becomes the only efficient solution (Dessus et al., 1992, p. 70; Fischer and Schrattenholzer, 2001, pp. 151–159; Yamamoto et al., 1999, pp. 101–113).

5.2 THE IMPORTANCE OF RENEWABLE ENERGY FOR THE ECONOMY

Renewable energy is widely regarded as one of the most effective solutions to improve the environment and mitigate and reduce greenhouse gas emissions (Socolow, 1991, pp. 121–126). Renewable energy also has a positive impact on economic growth and contributes to the development of countries (Inglesi-Lotz, 2016, pp. 58–63).

Conditions for the Development of Renewable and Civic Energy 89

TABLE 5.2

Socio-economic Effects of Renewable Energy Use

Macroeconomic effects		Distributional effects		Effects related to the energy system		Additional effects	
Gross impact	Net impact	Positive	Negative	Positive	Negative	Benefits	Costs
- Gross domestic product, - Prosperity, - Employment, - Trade balance.		- Ownership structure - Energy distribution channels, - The impact of energy distribution on energy consumers and taxpayers.		- Additional costs for energy generation and balancing, - The additional costs of network transmission and transactions, - Externalities.		- Risk reduction, - Developing entrepreneurial and civic-minded attitudes, - Environmental education, - External effects.	

Source: Own elaboration adapted from IRENA (2016, p. 44).

Renewable energy has a significant impact on the socio-economic development of the world and individual regions. The socio-economic effects of this impact can be divided into four categories: macroeconomic, distributional, energy system-related and other (see Table 5.2).

Initiatives in renewable energy are always undertaken under certain specific conditions and times, the consequence of which should be an individual approach to each implementation. There are a number of issues that should be considered and addressed prior to investment. Issues related to renewable energy in modern economies can be considered in the technical, organisational, economic, social, legal or environmental areas. (see Table 5.3).

Renewable energy has the potential to impact the economy on both global and regional/local scales. A 100% increase in the share of renewables in the global energy mix by 2030 would result in an increase in global GDP of 0.6% or approximately 700 billion USD. Most of these positive impacts on GDP would be due to new investments in renewables, which will have economy-wide effects (IRENA, 2016, p. 44). The aforementioned positive developments are confirmed by the Global Energy Transformation Report, which presented forecasts that GDP in the global economy is expected to increase between 2018 and 2050 due to RES investments (a cumulative gain of 52 trillion USD) (IRENA, 2018, p. 47).

The development of renewable energy is increasing the demand for coupled equipment and services, as well as fossil fuels. Trade in renewable energy installations and other goods and services will increase as a result of increased interest in the use of renewables in the energy sector and in buildings, industry or transport. This will also result in a decline in trade in industries related to other energy sources—including fossil fuels in particular (IRENA, 2016, p. 44). The availability of green technologies is also becoming increasingly important to many entities, which in many cases offers opportunities for competitive advantage or at least inclusion in ongoing business ventures (Dowell and Muthulingam, 2017, pp. 1288–1289). Increasingly important in modern economies for a company's contractors or customers is its zero-carbon performance, which can only be achieved through the use of RES (Fücks, 2016, pp. 185–193).

TABLE 5.3
Main Problem Areas Related to Renewable Energy

Area	Main problems
Technical	- Analysis of the possibilities of using and applying available technologies
	- Analysis of the impact of the investment on the electricity system,
	- Selection of parameters of the generating equipment in line with available resources and capabilities of its own or the electricity system,
	- Assessing the feasibility of accepting energy into the distribution or transmission network from the installed generation capacity.
Organisational	- Identification of renewable energy resources and opportunities for their use,
	- Searching for areas for investment,
	- Assessment of energy disposal options.
Economic	- Raising the funds necessary for the implementation of investments,
	- Investment viability analysis,
	- Development of new services and impact on the local labour market,
	- Stimulating local entrepreneurship,
	- Changes in the income of local authorities, businesses or households.
Social	- Measuring and analysing the benefits associated with improved quality of life as a result of investments in renewable energy sources,
	- Shaping an environmentally friendly culture and attitudes,
	- Promoting new and innovative technologies,
	- Networking,
	- Public participation in the system—with particular reference to citizen energy,
	- Providing a sense of security.
Legal	- Assessment of the feasibility of connecting the unit under non-energy conditions (land use, environmental protection, property rights, etc.),
	- Compliance with regulations and connection conditions, concessions, tariffs, etc.,
	- Obtaining all types of permits (building, operating, etc.),
	- Meeting safety standards and guidelines for RES development,
	- Setting standards for power sale agreements, considering the efficiency of power plant operation and the conditions for power reserve.
Environmental	- Measuring and analysing the environmental benefits and provision of clean energy,
	- Issues of changes in the use and management of environmental resources,
	- Spatial planning for investment,
	- Changes in landscape and space associated with renewable energy investments.

Source: Own elaboration.

Increasing the share of renewable energy in the global energy system also completely changes trade relations worldwide. These changes will affect both importers and exporters of non-renewable raw materials. For countries that import fossil fuels, increasing the share of renewable energy may have favourable trade implications through improved balances of payments and increased energy security. Fossil fuel exporters, on the other hand, will be in a much more difficult situation, where dependence on export revenues can have a significant impact on their economies. In addition, these are also very often politically and socially unstable countries, where

Conditions for the Development of Renewable and Civic Energy

a disruption of the current situation can contribute to increased social and political unrest and tensions, both locally and globally. This problem has been extensively described in the literature and commonly operates under the term 'paradox of abundance' or 'curse of natural wealth' (Gylfason, 2001, pp. 847–855; Rosser, 2006, pp. 1–36; Sachs and Warner, 1999, pp. 43–76; Torvik, 2002, pp. 455–470). On the other hand, however, the early switch of these countries to renewable energy production can be seen as an opportunity to diversify the economy and thus create new development opportunities.

In addition, especially in households that use their own heating systems (e.g. coal boilers), by replacing them with renewable energy sources, the need for users to spend time to operate them is drastically reduced. Being able to devote some of this time to relaxation, family bonding or personal development can bring many social benefits.

The development of energy systems cannot be considered in isolation from energy poverty issues. The development of renewable energy, especially of a dispersed nature, through its local character and versatile applicability, can contribute to alleviating energy poverty. On the other hand, when households are not able to use the simplest, non-organic resource-based energy generation systems and are faced with the need to incur significant capital expenditure for renewable energy installations, the problem of energy poverty may be exacerbated. This applies both locally and regionally, as well as globally.

The development of decentralised energy based on renewable energy sources is also changing the direction of the cash flow resulting from the energy payments. In the case of fossil fuel use, these funds flow out of the region contributing to the wealth of other communities (e.g. regions with power stations or coal mines in their area). By contrast, when renewable energy sources are used, the money stays in the area, providing an additional source of income for local communities. The establishment of new businesses (production and operation of installations) and the increased economic activity of the region's inhabitants (e.g. biomass production) additionally contribute to an increase in income from taxes or local levies.

Today, globally, nearly two billion people have no access to commercial forms of energy, and billions more have only short-term and unstable access. The continuation of this state of affairs threatens to seriously jeopardise the   maintenance of stability and quality of life worldwide (World Energy Council, 2007, p. 72). Creating an effective model for renewable energy development, of which civic energy is an essential component, is also part of the fight against poverty. If it is acknowledged that one of the reasons for the persistence of poverty is the lack of exchangeable ownership or capital (Kotler, 2016, p. 29), the renewable energy model should reduce the barriers of entry into the system, in such a way that everyone involved can become a shareholder or co-owner of the venture.

The impact of renewable energy technologies on creation of new jobs is one important argument in favour of renewable energy sources (Sidorczuk-Pietraszko, 2015, pp. 26–41; Goldemberg, 2006, p. 8). According to estimates made by the International Renewable Energy Agency, there were (directly or indirectly) 12 million people working in the renewable energy sector in 2020.12 million people. This number has continued to grow worldwide over the past decade. Clearly, there is huge potential for growth, and according to the analysis of the quoted Agency, employment in the

RES in 2030 is expected to remain concentrated in the technologies used today. Most renewable energy employment would be in bioenergy (9.1 million jobs by 2030), solar (8.5 million jobs by 2030), hydropower (3.8 million jobs by 2030) and wind (2.2 million jobs by 2030) (IRENA, 2018, p. 51). It is important to mention here that the employment growth in renewable energy will mainly take place in highly developed countries, while in underdeveloped countries (mainly African countries), jobs in renewable energy will not offset the loss of jobs in the coal sector (Garcia-Casals et al., 2019, pp. 113–114). Global renewable energy employment by technology in 2020 (thousands) is presented in Table 5.4.

The impact of renewable energy sources on job creation is one of the main arguments raised in favour of renewable energy support policies. Numerous studies (Gostomczyk, 2013, pp. 122–127) on direct employment in the renewable energy sector clearly indicate that the renewable energy sector is responsible for job creation, both in the construction of installations and in the exploitation or maintenance phases (see Table 5.5). Also, these jobs are local, that is, created in regions where RES installations are located.

Both in the public debate and in media materials, the extent of support for renewable energy is highlighted. From a technical point of view, this is, of course, a very simple procedure, which is determined by the transparency of support mechanisms and the ease of determining the directions and costs of intervention. Such a situation unequivocally creates a negative image, attractiveness and usefulness of renewable energy for the process of social and economic development of particular regions or the country. For it should be certainly stated that each type of power industry has benefited or benefits from public funds (e.g. conventional power industry, based on coal, often receives support through indirect forms, which are difficult to distinguish and value, including environmental or opportunity costs).

A fundamental characteristic of renewable energy is its innovativeness (Johnstone et al., 2010, pp. 133–155; Nesta et al., 2014, pp. 396–411; Richter, 2013, pp. 1226–1237).

TABLE 5.4
Global Renewable Energy Employment by Technology in 2020 (Thousands)

Specification	Number of employees
Solar photovoltaic	3,975
Liquid biofuels	2,411
Hydropower	2,182
Wind energy	1,254
Solar heating/cooling	819
Solid biomass	765
Biogas	339
Geothermal energy	96
Municipal and industrial waste	39
Others (tide, wave and ocean energy, and jobs not broken down by individual renewable energy technologies)	105
Total	12,017

Source: Own elaboration adapted from IRENA (2021, p. 20).

TABLE 5.5
Jobs Directly Created in Renewable Energy During Construction and Installation, As Well As Exploitation and Maintenance Phases

Energy source	Unit	Construction and installation	Exploitation and maintenance
Wind energy	[MW]	13.0	0.20
Solar thermal energy	[1,000 m²]	2.5	5.00
Solar energy—photovoltaic	[MWp]	34.6	2.70
Biofuels	[1,000 t/year].	5.0	1.50
Hydropower	[MW]	18.6	1.40
Biomass—heat	[tep].	0.1	0.01
Biomass—electricity	[MW]	4.0	0.14
Biogas	[MW]	25.0	6.00

Source: Own elaboration adapted from Sidorczuk-Pietraszko (2015, pp. 28–29).

When analysing changes in the renewable energy sector, innovation should also be looked at very broadly, considering both technology-related changes and non-technological factors. The development of the renewable energy sector can therefore take place both through technological (technology change) or non-technological innovations (change in the organisation of the energy system, the way energy is financed, the environmental awareness of society, etc.) (Bigliardi and Dormio, 2009, p. 224). P. Kotler, like many other researchers, identifies renewable energy both as a potential source of great innovation and as a place where many jobs will be created in the future (Kotler, 2016, p. 114). The socio-economic changes brought about by the development of renewable energy in the literature (Fücks, 2016, p. 197–199; Swilling, 2013, pp. 14–22; Weizsacker, 2009, pp. 12–15) are often referred to as key innovations that are the start of another, this time 'green' Kondratiev cycle (Bartkowiak, 2000, pp. 19–25). Kondratiev's new green cycle is achieved through key innovations: increased resource efficiency, renewable energy, system design and biomimicry (bionics) (Fücks, 2016, pp. 202–203).

5.3 THE IMPACT OF RENEWABLE ENERGY ON THE NATURAL ENVIRONMENT AND HUMAN HEALTH

The dominant share of conventional power generation and the pollution emitted as a result of its operation clearly demonstrate that its negative environmental and health impacts are undeniable, and that maintaining the current state, or worse, giving permission to its deterioration, will have far-reaching negative consequences (McMichael, 1991, pp. 499–501).

The main reason for the switch to renewable energy sources is therefore the enormous pollution generated by the combustion of fossil fuels (mainly coal and oil). Consequently, energy production based on non-renewable raw materials is increasingly being shifted away from fossil fuels and is generally being replaced by renewable sources, which are widely regarded as clean and allowing for an optimum use of resources, consequently minimising environmental impact. Furthermore, energy production from renewable resources generates minimal secondary waste.

The transition from conventional to renewable energy sources provides significant public health benefits. Air and water pollution emitted by coal-, gas- or oil-based facilities carries negative consequences related to respiratory problems, neurological damage, heart attacks or cancer (Brunekreef and Holgate, 2002, pp. 1233–1242; Kampa and Castanas, 2008, pp. 362–367; Review of evidence on health aspects of air pollution—REVIHAAP Project, World Health Organization Europe, 2013, pp. 4–25). Replacing fossil fuels with renewable energy reduces premature mortality and loss of working days, as well as overall health care costs. It is worth citing here, for example, the results of studies carried out by international research centres, whose research applies to the economy as a whole, commercial energy or household solutions.

According to calculations by HEAL, air pollution caused by the combustion of coal in power stations alone causes annual costs of illness and death in the EU of between €15.5–42.8 billion. These costs are covered by a variety of sources, ranging from healthcare budgets, the economy in general (e.g. loss of productivity) and individual household budgets and savings (The Unpaid Health Bill. HEAL, 2013, pp. 25–26).

There is also an inextricable link between conventional and nuclear energy and water. It is estimated that, currently, about 15% of the world's extracted freshwater is used for energy purposes (IRENA, 2016, p. 45), and in industrialised countries, the energy sector accounts for about 40% of freshwater consumption. The problem of water demand from the energy sector can be crucial for the socio-economic development of a country (Byers et al., 2014, p. 16). A switch to renewable energy could reduce this consumption by almost half in the UK and by more than a quarter in the US, Germany and Australia (IRENA, 2016, p. 45). Given the huge demand for water within Polish energy sector, the degree of water demand reduction in the event of an energy transition and a switch to RES, on a national scale, could be even greater.

When considering the environmental impact of renewable energy sources, their positive character should be clearly emphasised (Panwar et al., 2011, pp. 1513–152). Any negative factors that may arise during the installation production phase or in the production process itself are usually short-term and reversible.

Like conventional or nuclear power, the renewable energy sector also consumes energy at various stages of the infrastructure life cycle—from the extraction of the necessary raw materials, through component manufacturing and plant assembly, to the decommissioning and clean-up of the technology. One indicator describing energy intensity in this respect is the energy payback time (EPT). The value of EPT depends on a number of factors, of which the degree of sophistication of the technology used to produce the infrastructure and installation is undoubtedly a key one. As can be seen from the analysis of the studies (see Table 5.6), renewable energy achieves similar indicators and, under favourable climate conditions, being able to operate at high efficiency, much better results than conventional energy.

However, it sometimes happens, especially in the case of photovoltaic installations or CSP systems, that the EPT indicator assumes very unfavourable proportions. Such situations are usually observed in regions with limited sunshine, where weather conditions prevent the full uptake of energy. As a consequence, this leads to solar energy becoming less competitive than conventional or nuclear technologies.

TABLE 5.6
Energy Payback Time (EPT) by Technology

Technology/resources	Energy Payback Time (EPT)	
	From	To
Lignite	1.9	3.7
Natural gas, open cycle	1.9	3.9
Natural gas, combined cycle	1.2	3.6
Heavy water reactor	2.4	2.6
Light water reactor	0.8	3.0
Photovoltaics	0.2	8.0
Geothermal	0.6	3.6
Wind energy	0.1	1.5
Hydroelectric power plants	0.1	3.5

Source: Own elaboration adapted from Renewable energy sources and climate change mitigation. Summary for policymakers and technical summary. Special report of the intergovernmental panel on climate change. Intergovernmental Panel on Climate Change, Potsdam, 2012, p. 731.

Significant controversy is also caused by the need to secure large areas needed for the installations using renewable energy sources (e.g. wind power plants or photovoltaic farms), or the use of land for the supply of raw materials for renewable energy (especially biomass production). In the first case, costs are incurred due to the reduction in natural flora (where present) at the location of the equipment and associated facilities. This is a clearly negative, direct and permanent impact. Like any infrastructural investment, the investment in renewable energy sources also clearly affects the process of landscape transformation via introduction of an anthropogenic element of an industrial character. In addition, electromagnetic radiation or noise emissions are still present in many respects, and due to the specificity of plant operation, there is a constant impact. In the second case, there may be problems associated with increasing the area of vegetation, which in extreme cases can represent a certain barrier to proper air circulation. Potential risks may also arise in the case of excessive chemicalisation with plant protection products and artificial fertilisers.

5.4 INFLUENCE OF SOCIAL AWARENESS ON THE POSSIBILITIES AND DIRECTIONS OF DEVELOPMENT AND USE OF RENEWABLE ENERGY SOURCES

The processes related to socio-economic development taking place in the modern world led to the development of social awareness, which not only transforms with the systemic changes but is also an important factor influencing the transformations taking place. In practice, this means that social awareness plays a modifying role, which may prove to be a beneficial factor, or one that inhibits or even prevents change (Maciejewski, 2009, p. 8). The formation of social awareness and, consequently, social attitudes is influenced by systemic changes that alter the current way of acting or behaving. They are particularly problematic when they are introduced in a violent and revolutionary manner.

Due to the innovative nature of renewable energy sources, the possibility to introduce rapid changes when making an energy transition, for obvious reasons met with different public concerns and expectations.

The social aspects associated with renewable energy can be grouped into five categories (Szulecki and Szwed, 2013, pp. 190–191):

- The spatial organisation of the energy system (referring to the spatial location of renewable energy industries, characterised by their dispersed nature and aiming to ensure a balanced level of consumption in a given area). It refers to the acceptance of individual installations and their spatial intrusion.
- Governance structure (concerns the institutional, political, and administrative organisation of the renewable energy sector). A key issue of this category is the problem of legal, administrative or ownership authority over renewable energy industry and the possibility of cooperation with conventional energy industry.
- Political economy (a category dealing with the distribution of benefits and costs of renewable energy). It considers issues related to individual and group investment decisions, construction of new economic initiatives or the impact of renewable energy on the labour market.
- Ecological strategy (concerns changing the way communities interact with the environment through the use of renewable energy sources). This category considers the relationship between society and environment, as well as the contribution of renewable energy to the possibility of combating climate change and ensuring environmental security.
- Energy security in the broad sense (concerns the impact of renewable energy on the quality and standard of life through the possibility to provide secure energy for society and its needs). Another important issue is the impact of renewable energy sources on the possibility of achieving energy self-sufficiency and making society and its economy independent of export of energy raw materials or energy.

The global sociological literature points out that energy rarely breaks through into the consciousness of societies or politicians (Rosiek, 2012, p. 200). This has especially been the case in recent years, where the price of energy has been relatively widely accepted socially. Low awareness of the benefits that the economy and society would gain by basing their production on renewable energy still prevails among the public. As can be seen from studies on the social acceptance of specific energy sources, or preferences for specific types of installations, renewables are rated strongly positively, and coal much less so (nuclear fuel is also rated very low) (Kardooni et al., 2018, pp. 659–668; Kim et al., 2018, pp. 761–770; Nel et al., 2016, pp. 62–70, Ruszkowski, 2015, pp. 69–73; Wüstenhagen et al., 2007, pp. 2683–2691).

When analysing investment projects in renewable energy, as for the entire energy sector, NIMBY effects should be identified and recognised (see Renewable Energy and the Public. From NIMBY to Participation, ed. P. Devine-Wright, Routledge, 2010). As the literature analysis and case studies show, NIMBY effects in renewable energy are much milder in nature than, for example, in nuclear or conventional energy.

Conditions for the Development of Renewable and Civic Energy

In the case of RES, they are substantially related to wind power or the construction of waste incineration plants. Concerns in these cases are usually related to alleged health hazards, noise concerns and landscape change, which can consequently lead to a decrease in quality of life and property values (Horst, 2007, pp. 2705–2714; Sims et al., 2008, pp. 251–269). Overcoming the effect can be done through education or by inviting individual stakeholders to cooperate—and sharing the profits made from completed projects.

The energy transition requires that any changes in national energy policy must be preceded by improvements in public awareness. Breaking down the knowledge barrier becomes crucial. An excellent and widely cited example of how the public can be involved in the process of building energy systems is that of Denmark, where through education and inclusion of public participation in energy policy, incredible success has been achieved and a competitive and socially attractive energy system has been built (Lund, 2000, pp. 249–259; Lund and Mathiesen, 2009, pp. 524–531; Mendonça et al., 2009, pp. 379–398).

5.5 DIRECTIONS AND TRENDS IN THE DEVELOPMENT OF RENEWABLE ENERGY—THE KEY ROLE OF CIVIC ENERGY

Globally, the potential for renewable energy is more than sufficient (Vries et al., 2007, pp. 2590–2610; Moriarty and Honnery, 2012, p. 247; Resch et al., 2008, pp. 4048–4056), and exceeds energy demand many times (even at the current technical level). Geothermal energy and solar energy have the greatest technical and theoretical potential (see Table 5.7).

The share of renewable energy in global energy consumption in 2020 was 12.6%. The estimated share of energy in global energy consumption in 2020 is presented in Figure 5.1.

An analysis of financial data on investments in renewable energy shows that they are significantly higher globally than investments in conventional and nuclear power. Reference should be made, for example, to the REN21 report, which states that RES continued to attract far more investment in 2021 than did fossil fuel-based or nuclear generating plants (investment in RES power capacity accounted for 69% of the total investment committed to new power generating capacity) (Renewables, 2022 Global Status Report. REN21, 2022, p. 181).

TABLE 5.7
Global Potential of Renewable Energy Resources (in EJ)

Resources	Technical capacity	Theoretical potential
Hydroelectric power stations	50	150
Biomass	> 250	2,900
Solar energy	> 1,600	3,900,000
Wind energy	600	6,000
Geothermal energy	5,000	140,000,000
Ocean energy		7,400
Total	> 7,500	143,000,000

Source: Own elaboration adapted from Johansson et al. (2004, p. 3).

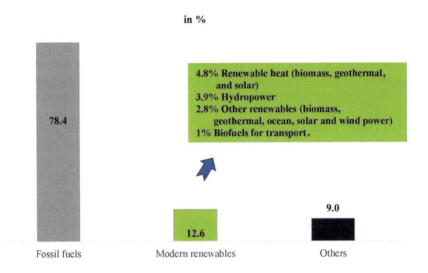

FIGURE 5.1 Share of Modern Renewable Energy in 2020 (in %).

Source: Own elaboration adapted from Renewables, 2022 Global Status Report. REN21, 2022, p. 37.

Of interest in terms of new investment in renewable energy is the fact that global investment in renewable energy reached a record-breaking level in 2021, estimated at 366 billion USD. This was an increase of 6.8% compared to 2020, mainly due to the global growth of photovoltaics (PV). Investment in renewable energy has already exceeded 250 billion USD per year since 2014 (Renewables, 2022 Global Status Report. REN21, 2022, p. 175). Global Investment in Renewable Power and Fuels for the period 2011–2021 is presented in Figure 5.2.

With current trends in the development of renewable energy, it is estimated that by 2030 renewable energy sources will account for approximately 50% of total energy production (world rated power). The ongoing research work and observed trends of increasing interest and use of renewable energy worldwide are still not sufficient to supply the world with 100% renewable energy by 2050. A similar perspective is presented in the BP Energy Outlook, where it is estimated that RES will dominate new energy investments in the next 20 years, even becoming significantly cheaper than coal-based power plants. According to the report's authors, within two decades, assembled renewable energy installations will account for almost 50% of new power generation capacity. This will lead to a situation where the share of renewables in the electricity mix for Europe will increase to almost 40% by 2035, and for the United States of America or China and the rest of the world will approach 20% (BP Energy—Outlook, 2017).

However, in order for this to happen, both support schemes and renewable energy regulations need to balance the interests of (often very particularistic) multiple parties. Assuming that the environment is the main stakeholder, the other stakeholders can instead be grouped into homogeneous groups that have

Conditions for the Development of Renewable and Civic Energy

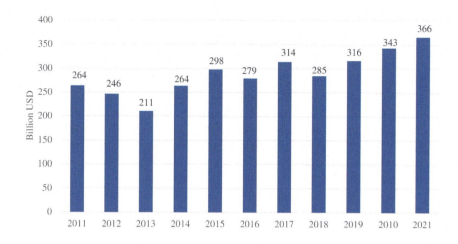

FIGURE 5.2 Global Investment in Renewable Power and Fuels, 2011–2021.

Source: Own elaboration adapted from Renewables, 2022 Global Status Report. REN21, 2022, p. 176.

and will continue to have a significant impact on RES development and use. These primarily include:

- **Project promoters.** This group should include all those who initiate and organise projects related to the use of renewable energy sources.
- **Investors.** Investors are entities that commit capital or other resources to a renewable energy project in order to receive a satisfactory return on their investment. Investors in the renewable energy industry can range from companies, banks, private equity and venture capital funds, pension funds, corporations, housing associations, cooperatives, public sector bodies and religious associations to foundations, associations and individual private consumers. Each of these investor groups is characterised by different expectations in terms of financial return on investment, different levels of risk acceptance, the time over which they can commit capital, and different criteria for assessing attractiveness. Therefore, the investor support schemes put in place can be targeted at all groups or selected groups of investors. Despite the differences between the different groups, most investors consider the payback period, risks and environmental issues when considering investment decisions.
- **Manufacturers.** A stakeholder group comprising manufacturers of renewable energy installations and other necessary equipment. In a broader sense, this group includes all companies involved in the supply chain from the design and production of renewable energy installation components to companies involved in installation and service work. Manufacturers are a group that also often invests in research and development, construction of production and service facilities and staff training.

The main determinant of the group's commitment to further projects is the prospect of stable and predictable long-term growth in demand for particular renewable energy technologies and equipment.

- **Consumers.** Consumers are the stakeholder group that has a direct interest in the need to secure their own energy needs. For reasons of self-interest, they are primarily interested in the possibility of supplying energy at the lowest cost, while maintaining continuity and security of its use. Other factors determining the acceptance of energy sources are also important, such as the impact on the environment (mainly human health), the possibility of generating additional revenue or other externalities.

Consumers constitute the largest group of stakeholders, who are located in different regions characterised by different social, economic and historical conditions in terms of energy consumption demand and culture. The dispersed nature of consumers also makes them a special group for renewable energy. Indeed, almost every consumer may not only be an energy consumer but also be an energy producer.

- **Public authorities.** This group has a significant impact on other stakeholder groups and has a direct influence on energy policy and, consequently, on the shape of the energy mix. The energy system of each country and region is an important element of public policy, which defines strategic issues such as the volume and source of production, investment priorities or methods of energy distribution. Through public policy, energy demand and consumption levels can also be significantly influenced. A number of tools and instruments can be used in energy policy (e.g. international agreements and commitments, national legislation at central, regional and local level, investment incentives, conservation and energy efficiency guidelines, taxes, penalties and other instruments (see Table 5.8).

Public authorities also make direct decisions on investment projects in the energy sector. They make investments either on their own (e.g. in public buildings) or become the originator of projects with other entities (e.g. households or businesses).

The public sector can also support various entities through regulation, a system of subsidies or, for example, subventions. Both the regulations and the support given to the various groups, to be effective, must be adequate and should reflect the actual costs of renewable energy generation. If the system offers insufficient support or stifles renewable energy initiatives, the industry will not be able to develop. In an extreme situation, where the support is too high and regulations unilaterally favour renewable energy sources, the industry may become over-invested.

Although the cost comparison between renewables and conventional generation depends on the location and conditions of the country or region, the LCOE suggests that by using renewables it is already possible to offer energy services at a price that is very competitive with conventional energy. Furthermore, technological advances and greater efficiency are expected to further increase the price competitiveness of all types of RES compared to conventional technologies.

TABLE 5.8
Potential Instruments that Can Be Used by Public Authorities for Energy Policy Planning

Instrument	Instrument characteristics
Direct regulation	Orders and control (e.g. with regard to the imposition of energy production requirements based on renewable energy sources).
Direct public investment in research and development	Funding for research into the development of new technologies using renewable resources and the establishment of new research and development facilities.
Tax incentives and subsidies for those investing in research and development	Tax exemptions and subsidies for private entities investing in research and development of new technologies using renewable resources.
Market price instruments: excise duties, subsidies	Making market prices realistic to reflect externalities.
Feed-in tariffs (FIT)	Offering long-term contracts for renewable energy producers.
Information campaigns	Education, information campaigns, provision of advice on raising investment funds, technical solutions, consumption and behavioural patterns, as well as implementation of renewable energy investments.
Standardisation of products	Definition of minimum requirements to be met by products in the renewable energy industry.
Market performance standards	Introduce standard-based regulation of activities (e.g. obligation to source energy from RES or low-carbon fuel standards) Permits or certificates can be divested to entities that do not meet the relevant standards.
Transparent rules	Introducing requirements for companies to provide certain information about the services, products or capacities provided, terms and conditions of functioning.
Macroeconomic policy	Fiscal or monetary policy implemented to stabilise the economy and provide market liquidity and reduce the loss of credit score of entities operating in the economy.
Corporate taxation	Adjusting corporate income tax to a level that makes investment in renewable energy attractive and provides investment incentives for companies.
Competition policy/law	Competition and consumer protection activities including, inter alia, taking antitrust action.
Regulatory change	Eliminate regulatory errors and loopholes that make renewable energy more attractive.
Copyright	Copyright to provide incentive remuneration to inventor-innovators for further work.

Source: Own elaboration adapted from Gillingham and Sweeney (2010, p. 80).

- **Political parties and trade unions.** When representing the interests of their constituents or members, political parties and trade unions promote their own solutions to the energy system. Therefore, the interests they represent very often conflict with the interests of other entities or the broadly defined public good. Political parties or trade unions can become entities promoting and supporting the development of renewable energy, or, on the contrary, entities negating the legitimacy of renewable energy development and taking initiatives supporting other solutions (e.g. supporting coal-based energy).

- **Community organisations**. Social organisations are an important force in the development of renewable energy. They are becoming entities with influence on decisions taken by central or local administration but can also significantly create patterns of behaviour or consumption. Their influence on the process of educating society on pro-environmental attitudes is also substantial. Very often, social organisations also undertake investments in renewable energy, becoming an active player on the market.
- **Investment service providers**. Any production and service entities working for the renewable energy sector.
- **Education.** The functioning of the energy sector is significantly influenced by the education system. The way society is educated influences the level of consumption (including energy), energy efficiency or the use of innovative energy sources—including, among others, renewable energy sources.
- **Media.** Media have an important influence on the promotion of behavioural patterns and the popularisation of products and lifestyles. They currently play a particularly important role, not only as a source of information but also as an element of a complex system of social communication. The media, which are an indispensable element of modern humans' life, significantly influence the ongoing transformations in the intellectual, social and consumption spheres. Modern media are now recognised as the main source of knowledge, including that concerning environmental protection or energy, and are an important carrier for the transfer of knowledge concerning the use of technical solutions, equipment or pro-environmental behaviour.

All the aforementioned stakeholders operate in a network of interrelationships, and interactions. A network organisation is characterised by a permanent grouping of autonomous companies or other specialised entities that participate in cooperative links, both in market-based economic processes and in building social relationships. Furthermore, as noted by

M. Kilduff and W. Tsai (Kilduff and Tsai, 2013, p. 13), the network approach to diverse problems of today is becoming a new paradigm for development. With the need to adapt to ongoing changes in energy systems, the ability to create flexible networks between different entities and organisations is becoming increasingly important.

In the case of renewable energy, the problem is particularly important. This is determined, among other things, by the dispersed nature of renewable energy sources. Acting on the basis of the network, both producers and consumers of energy (as well as entities working for renewable energy) have the possibility to create extensive areas of influence, characterised by instability of forms and discontinuity of activities (Malara, 2006, p. 114). Thanks to such conditions, operating in a turbulent environment, they have the opportunity to adapt quickly to changes in the socio-economic and legal areas. The flexibility of networks, related to possibilities of rapid configuration changes, can take various forms, including those expressed in the

Conditions for the Development of Renewable and Civic Energy **103**

implementation of strategies related to knowledge management and organisational learning. Networking is also a way of transferring good practices, spreading innovation and creating own business ventures.

As can be seen from the considerations the development of renewable energy depends on many factors. On the one hand, economic conditions have an impact, on the other hand, political, social or environmental aspects do. Key factors influencing and conditioning the development of renewable energy include (Bilgili et al., 2015, pp. 323–334; Burger et al., 2014, p. 59):

- the pace of technical and technological development within renewable energy sources and energy efficiency,
- the costs of technologies using renewable energy sources,
- fossil fuel costs,
- the political and public dimension of support for renewable energy sources,
- acceptance of installations using renewable energy sources by local communities,
- an increase in demand for electricity, heat and fuel,
- integration of RES power generation facilities into existing infrastructure,
- integration of RES installations into energy markets,
- the development of decentralised energy systems,
- fashions and trends in lifestyle, consumption and ownership.

The production of electricity, heat or biofuels from renewable energy sources is becoming a priority in energy policy strategies at both national and global levels. When examining the prospects for the development of renewable energy sources in different regions of the world, the need for a detailed analysis of the specific conditions for a particular region or country becomes crucial. These conditions must always be considered in very broad terms and include, among other things:

- the potential of renewable energy sources,
- capital investment opportunities in the energy sector,
- preferences in the use of energy resources (political, social, etc.)
- historical conditions in the use of energy resources,
- public environmental awareness,
- the degree of creativity and innovation in the economy.

Dynamic changes will also take place in energy business models. Conventional energy will lose its monopoly, and power generation or distribution will become an ordinary business. Therefore, the key to success is the creation of new, innovative organisational solutions, enabling all aspects of socio-economic development to be holistically taken into account with technical and environmental capabilities.

All forms and methods of energy conversion and use entail some form of pollution and/or environmental degradation. Current energy systems in many countries, including highly developed ones, are heavily dependent on fossil fuels, which are concentrated in several regions of the world. Fossil fuels used for energy purposes contribute to air pollution on the one hand, and to the dependence of many countries on imports of energy resources on the other. For example, the EU's energy dependency was, in 2020, 57.5%. This means that the EU has to import more than half of the energy that is consumed within the community. It is worth noting at this point that over the last two decades (between 2000 and 2020), the energy dependency of the average EU27 country has increased from 56.3% to 57.5%, meaning that EU members have become slightly more dependent on energy imports. Among EU countries, Estonia was the least energy dependent in 2020 (10.6%). The highest energy dependency rates were recorded in Malta (91.6), Cyprus (93.1%) and Luxembourg (92.5%).

This situation should encourage individual countries to take initiatives to reduce the role of fossil fuels in energy systems and to limit possible supply disruptions and prevent uncontrollable changes in prices of energy resources. Improving the environment and reducing dependency is also increasingly becoming a priority in the economic policies of countries that do not accept the increasing costs associated with environmental protection, the purchase of fossil fuels and the associated risks (Goldemberg, 2006, p. 9). Renewable energy sources are expected to soon be suitable alternatives to conventional energy and their development will ensure that energy development will be balanced out in the future. These expectations are dictated by the fact that renewable energy sources (Turkenburg, 2000, pp. 220–272; Burger et al., 2014, p. 55):

- are a vast and inexhaustible source of energy,
- lead to the diversification of energy sources by increasing the share of various renewable sources, thereby increasing energy security,
- are widely available and thus reduce the geopolitical dependence of countries, as well as minimising the expenditure on imported fuels,
- contribute in most cases to a reduction in air pollution, thereby reducing damage to human health, animal kingdom and quality of the environment,
- characterised by the dispersed nature of installations (which in many cases can operate off-grid), contribute to improving access to energy services in rural and remote areas and contribute to the resilience of the electricity system,
- offset the consumption of fossil fuels and consequently saves them for other uses in the future,
- improve the development of local economies and create jobs,
- do not emit greenhouse gases into the atmosphere,
- reduce dependence on imported fuels,
- are an instrument for the environmental education of the public.

A key, important socio-economic benefit that can be achieved through the development of renewable energy is building of a citizen energy industry, based on local energy, environmental, capital and human resources. They are citizens and local communities, who can play a key role in a successful transition towards clean energy. Energy that, when produced and consumed locally, will allow for a more even redistribution of wealth, strengthen social justice and network of local cooperation.

6 Renewable Energy in the Polish Energy Sector—Resources and Use

CONTENTS

6.1 The Energy Sector in Poland 105
6.2 Resources, Potential and Use of Renewable Energy 111
    6.2.1 Solar Energy 113
    6.2.2 Hydropower 114
    6.2.3 Wind Energy 115
    6.2.4 Geothermal Energy 115
    6.2.5 Biomass.. 116
6.3 Legal Basis for the Use of Renewable
    Energy Sources in Poland in Terms
    of the Development of Civic Energy 119

6.1 THE ENERGY SECTOR IN POLAND

A number of unfavourable phenomena are intensifying in the world economy, forcing a new approach to energy policy, natural resources or the environment. The systematic increase in the price of energy resources, growing energy demand, failures of energy systems, increasing environmental pollution and political tensions, require a new approach to a broadly defined energy policy (Coady et al., 2015, pp. 8–13; Sovacool and Dworkin, 2014, pp. 88–318). Energy, being a strategic sector of any country's economy, significantly affects the socio-economic conditions of each country. It is also a sector of the economy with a detrimental impact on the environment and thus on human health. Therefore, it also directly affects the standard and quality of life of citizens (Bergstrom and Randall, 2016, pp. 3–74). This is particularly true for Poland, a country which, after the political changes that took place in the 1990s, has been faced with the need to transform its energy sector, which is currently based largely on conventional energy.

Analysis of statistical data shows that both total primary energy consumption and final energy consumption in Poland have remained at a similar level over the past years. In the case of total primary energy consumption, between 2010 and 2020, it remained in the range of 94.0 Mtoe (in 2014) to 105.9 Mtoe (in 2018). Final energy consumption between 2010 and 2020 remained in the range of 60.8 Mtoe (during 2014) to 73.8 Mtoe in 2018 (see Table 6.1).

DOI: 10.1201/9781003370352-7

105

106 Management of Civic Energy and the Green Transformation

TABLE 6.1

Energy Consumption (in Mtoe)

Specification	2010	2011	2012	2013	2014	2015	2016	2017	2018	2019	2020
Total primary energy consumption	100.5	101.5	98.1	97.7	94.0	95.2	99.5	104.1	105.9	103.5	101.8
Final energy consumption	65.6	64.2	63.7	62.4	60.8	61.4	65.7	69.7	73.8	73.0	70.5

Source: Own elaboration adapted from Energy efficiency in Poland in years 2010–2020, 2022.

TABLE 6.2

Production and Consumption of Primary Energy in the Years 2012–2020

Year	Production		Consumption	
	Total	Per capita	Total	Per capita
	PJ	GJ per capita	PJ	GJ per capita
2012	3038.9	78.8	4493.9	115.3
2013	3006.5	78.1	4488.3	116.6
2014	2854.6	74.2	4321.3	112.3
2015	2883.5	75.0	4430.3	115.3
2016	2804.2	73.0	4459.0	116.0
2017	2726.1	70.9	4480.4	116.6
2018	2613.9	68.1	4564.5	118.8
2019	2528.5	65.9	4438.6	115.6
2020	2377.8	62.1	4242.5	110.9

Source: Own elaboration adapted from Energy statistics in 2019 and 2020, 2021, p. 25.

Analysis of statistical data on primary energy in the national economy shows dynamic changes in both the amount of energy consumption in individual carriers and their importance in the overall structure. Obtaining of primary energy in Poland in 2020 amounted to 2,377.8 PJ. As can be seen from the statistical data presented in Table 6.2, obtaining of primary energy in 2012–2020, as well as obtaining and consumption per capita, systematically decreased.

Primary energy production in Poland uses mostly fossil fuels. The basic energy resource is coal (hard coal and lignite), which in 2020 had a 55.3% share in the structure of consumption of primary energy carriers in the national economy. The second most extracted carrier was lignite with a share of 15.5%. Natural gas accounted for 5.9%, oil for 1.7% and other, largely renewable energy sources for 21.6%. A decrease was recorded for hard coal, lignite and oil, while an increase was recorded for natural gas and other energy carriers. Energy obtaining per capita is above the European average in Poland (60.2 GJ in 2019) and amounted in 2019 to 65.4 GJ per capita. Among EU Member States, this placed Poland in the 11th place (Energy statistics in 2019 and 2020, 2021, p. 25).

Electricity is produced in Poland primarily in public thermal power plants. In 2020, the volume of production in these facilities amounted to 120 TWh, which

Renewable Energy in the Polish Energy Sector—Resources and Use 107

accounted for 76% of total production. The efficiency of public utility power plants has remained at a similar level for years and in 2020 was 42.6%. The most important fuel that was used to generate electricity was hard coal, with a share of 44%, and lignite with a share of 24%. Noteworthy, the share of these fuels in production decreased by 13 percentage points since 2014. Production from renewable energy sources accounted for about 18% and increased by 5 percentage points since 2014. The most important carriers in this group were wind energy, biomass and biogas. Solar energy has the smallest share but the highest growth rate. (Energy statistics in 2019 and 2020, 2021, pp. 28–29).

In 2021, the average annual power demand (with reference to the evening peak value) in Poland was 23,673.0 MW, while the maximum demand during daily peaks was 27,617.2 MW. The change that occurred compared to 2015 was 10%. The maximum domestic power demand in 2015 was on 7 January, in 2016 on 15 December, and in 2021 on 12 February. The minimum domestic power demand in 2015 was on 27 July, in 2016 on 15 August, and in 2021 on 6 June. In the case of the Polish electricity system, peak demand therefore occurs in the winter months and amounts to approximately 25,000–28,000 MW. The lowest electricity demand is approximately 11,000–13,000 MW, which occurs in the summer months. Such a large scale of demand variability requires generating units capable of starting and stopping operation and changing the generated power according to TSO instructions (see Table 6.3).

As can be seen from the data in Table 6.4, approximately 16 million households were electricity consumers in 2020. Approximately two-thirds of them were located

TABLE 6.3
Selected Data on the Operation of NPS from 2015 to 2021

Specification	Value (MW)		
	2015	2016	2021
Rated capacity of domestic power plants[1]	38,891.3	40,491.1	50,714.6
Available capacity of domestic power plants[1]	26,763.20	28,104.80	29,197.4
Power requirement[1]	22,218.60	22,482.60	23,673.0
Maximum national power demand	25,101.10	25,546.30	27,617.2
	7.01.2015 y.	15.12.2016 y.	12.02.2021 y.
	5:00 pm	5:00 pm	10:45 am
Power reserve on the day on which the maximum national power demand occurred[2]	3,441.10	3,637.70	4,257.2
Minimum national power requirement	12,650.30	11,276.80	12,132.7
	27.07.2015 y.	15.08.2016 y.	6.06.2021 y.
	4:45 am	6:00 am	5:00 am
Power reserve on the day on which the national minimum power demand occurred	11,049.30	12,921.10	13,620.7

1. Data based on annual average values from the evening peak.
2. Power reserve = spinning reserve in thermal centrally dispatched generating units + water centrally dispatched generating unit reserve + cold reserve in thermal centrally dispatched generating units.

Source: Own elaboration adapted from Activity Report. The President of the Energy Regulatory Office in Poland, 2017, p. 96; Activity Report. The President of the Energy Regulatory Office in Poland, 2022, p. 146.

TABLE 6.4
Household Electricity Consumers and Consumption

Specification	2010	2015	2018	2019	2020
Consumers in thousands	14,179	14,468	15,398	15,588	15,799
Cities	9,409	9,592	10,244	10,399	10,556
Electricity consumption per capita in kWh	773	736	794	798	822
Cities	785	728	777	778	804

Source: Own elaboration adapted from Housing Economy and Municipal Infrastructure in 2020, GUS, 2021, p. 46.

TABLE 6.5
District Heating Infrastructure and Sales of Thermal Energy

Specification	2021
Total thermal network in km (as of 31 December)	25,238.7
Cities	24,167.7
Rural areas	1,071.0
Thermal transmission and distribution network in km (as of 31 December)	16,524.8
Sales of thermal energy in 1,000 TJ (annually)	211.7
cities	208.7
rural areas	3.0
Sales of thermal energy to residential buildings in 1,000 TJ (annually)	166.5
Cities	164.6
Rural areas	1.9

Source: Own elaboration adapted from Energy management and gas supply system in Poland in 2021, GUS, 2022, p. 2.

in cities. Electricity consumption per capita in 2020 was 822 kWh and has been steadily increasing for many years.

Heat production in 2020 amounted to 285.9 PJ. Utility power plants, district heating plants and district heating boilers of the utility power industry accounted for the largest share of heat production. Heat production from industrial power plants and non-commercial district heating plants, as well as others, was much smaller with a total of 22.2 PJ in 2020. At the end of 2021, the length of the district heating network in Poland was 25, 238.7 km, of which 95.8% was located in cities (see Table 6.5). In 2021, 212,000 TJ of thermal energy was sold, of which 1,667,000 TJ (79%) was used for heating of residential buildings. Approximately 209,000 TJ (99%) of thermal energy was sold to urban residents, of which approximately 165,000 TJ was sold for heating residential buildings. In 2021, solid fuels (62%) and gas fuels (36%) accounted for the largest share in the structure of fuels used to produce thermal energy for heating, while the least thermal energy for heating was produced using oil fuels (2%) (Energy management and gas supply system in Poland in 2021, GUS, 2022, pp. 2–3).

Poland is one of the European Union countries that are least dependent on energy imports. This is mainly due to solid fuels, where domestic demand was met by imports in 2019 in 6%. In the case of the next two significant carriers—natural gas and oil together with petroleum products—Poland's dependence on imports is slightly higher than the EU average: in case of oil, it amounted to 97% and in case of natural gas, to 83%. For the European Union, these values were 89% and 83%, respectively in 2019 (Energy statistics in 2019 and 2020, 2021, p. 35).

When analysing conditions for development of the Polish energy sector, it should also be kept in mind that in prospect of the coming years, a systematic decline in production of thermal coal from the currently operating mines may lead to creation of a so-called 'carbon gap'. The carbon gap, understood as a difference between the current level of electricity production from hard coal and lignite and the level resulting from a decline in production of these raw materials in the existing mines over the next decades, may also force solutions seeking a safer and more secure way of obtaining energy. Possible courses of action in this regard are:

- support for the construction of new deep mines and open pits and, where possible, launching of prospective deposits,
- increased imports of hard coal,
- development of gas technology based on imports of liquefied natural gas (LNG),
- development of nuclear power,
- promotion and development of RES—including mainly civic energy.

For operation of the energy sector, the issue of energy efficiency is also becoming crucial. Efficiency in the use of energy in the economy is an important factor affecting the costs of production and service activities, the profits made, the competitiveness of production, the social costs of maintaining infrastructure and the standard of living of the population. Inefficient energy use results in over-consumption of energy resources and environmental pollution problems.

The issue of energy intensity and, consequently, energy efficiency of the Polish economy becomes especially important in the perspective of projected demand for energy and directions of changes in energy consumption. In Poland, energy consumption does not increase in line with economic growth, which means that modernisation of the existing energy infrastructure and implementation of green technologies introduced, inter alia, as a result of the impact of legal and economic instruments, result in a decrease in energy consumption.

Interesting information about the energy intensity of the Polish economy is provided by the ODEX energy efficiency indicator. The ODEX indicator, which illustrates progress in improving energy intensity in relation to the base year (2000), is calculated for each year as the quotient of actual energy consumption in the given year and the theoretical energy consumption excluding the unit consumption effect (i.e. assuming the existing energy intensity of production processes of particular products).

As can be seen from the analysis of the ODEX indicator values (see Table 6.6), there is a downward trend in Poland. By 2020, it decreased both for households (to 77.0) and for transport (to 69.9). The largest decreases were observed in industry,

TABLE 6.6

ODEX Indicator, Base Year 2000 = 100

Specification	2010	2011	2012	2013	2014	2015	2016	2017	2018	2019	2020
Industry	54.0	52.4	51.6	51.0	50.2	49.4	48.8	48.3	47.9	47.3	47.1
Transport	88.3	86.1	81.8	77.4	74.7	74.5	74.5	74.3	73.7	71.2	69.9
Households	85.5	85.1	84.6	83.9	82.1	80.4	78.7	78.1	77.5	77.2	77.0
Total	75.4	73.8	72.3	70.6	69.1	68.0	67.3	66.7	66.0	65.3	64.8

Source: Own elaboration adapted from Energy efficiency in Poland in years 2010–2020, 2022, p. 63.

TABLE 6.7

Energy Efficiency Indicators in Households Sector in 2010, 2015 and 2020

Specification	Unit of measurement	2010	2015	2020
Energy consumption per dwelling	toe/dw.	1.621	1.336	1.388
Total consumption per m²	kgoe/m²	22.4	18.1	18.6
Heating consumption per m²	kgoe/m²	15.8	11.9	12.3
Electricity consumption per dwelling	kWh/dw.	2124.3	2001.7	1998.4

Source: Own elaboration adapted from Energy efficiency in Poland in years 2010–2020, 2022, p. 22–25.

where it reached a value of 47.1 in 2020, which means an increase in energy efficiency in the analysed sector by as much as 53%. Changes in the value of the ODEX indicator between 2010 and 2020 are presented in Table 6.6.

Improvements in energy efficiency in Polish households can be seen in both electricity and heat consumption. The statistics in Table 6.7 clearly indicate that household electricity consumption in Poland steadily declines. Household electricity consumption per dwelling in 2020 was 1998.4 kWh/apartment and was 5.9% lower than in 2010, when it was 2124.3 kWh/apartment. In contrast, thermal energy consumption per dwelling or per m² decreased noticeably between 2010 and 2015 and remained constant between 2015 and 2020.

Significant discrepancies in energy intensity of individual economies, whether national, regional or local, are influenced by a number of factors, the most significant of which include (Maśloch, 2018, pp. 257–258):

- geographical location, weather conditions and availability of natural resources,
- provision of energy, transport and housing infrastructure,
- human capital and lifestyle,
- available capital and access to aid programmes and funds to support energy investments,
- interest in and acceptance of renewable energy solutions,
- innovation of the economy (enterprises) and the provision of R&D facilities,
- local, regional and national historical and political conditions.

Renewable Energy in the Polish Energy Sector—Resources and Use 111

TABLE 6.8

Impact of Selected Factors on Variation of Final Energy Consumption in the Years 2010–2020 (Mtoe)

Specification	Industry	Households	Transport	Services	Agriculture	Total
Variation of consumption	2.8	−1.0	4.6	−1.6	0.1	4.8
Factors						
Activity	5.4	–	5.7	2.7	−0.1	13.7
Dwellings	–	2.4	–	–	–	2.4
Lifestyle	–	0.8	–	–	–	0.8
Structural changes	−1.2	–	1.7	–	–	1.9
Energy savings	−2.2	−2.2	−5.1	0.0	0.0	−9.5
Weather conditions	–	−3.5	–	−1.3	0.0	−2.2
Others	0.8	−1.7	2.0	−3.0	0.2	1.6

Source: Own elaboration adapted from Energy efficiency in Poland in years 2010–2020, 2022, p. 30.

The factors influencing change in energy consumption can be classified according to homogeneous groups, such as activity of the entities, number of dwellings, lifestyle, structural changes, energy savings and weather conditions (see Table 6.8).

As can be seen from analysis of the data in Table 6.8, the biggest influence on the change in final energy consumption was economic activity, an increase in which contributed to an increase in energy demand of 5.4 Mtoe for industry, 5.7 Mtoe for transport, 2.7 Mtoe for services, while decreasing demand by 0.1 Mtoe for agriculture. In households, the factors driving the increase in energy demand were an increase in the number of dwellings and a change in lifestyle (larger dwellings). Structural changes in industry reduced energy consumption by 1.2 Mtoe, while in case of transport they increased the consumption by 1.9 Mtoe. Energy savings reached a total of 9.5 Mtoe, with the largest savings achieved in transport (5.1 Mtoe). Weather conditions reduced the energy consumption by 4.8 Mtoe and other factors increased the consumption by 1.6 Mtoe (Energy efficiency in Poland in years 2010–2020, 2022, p. 30).

6.2 RESOURCES, POTENTIAL AND USE OF RENEWABLE ENERGY

The amount of renewable energy sources depends on natural conditions, geographical location and the state of agriculture and forestry. Therefore, the level of renewable energy sources and their availability varies throughout the country. It should also be kept in mind that the potential resources, given the current technical, economic and legal conditions, are not fully recoverable. In the literature on renewable energy, different opinions can be found on renewable energy sources in Poland (Maśloch, 2012, p. 51).

Renewable energy sources, apart from hydroelectric and geothermal power, which resources are well recognised, have not been accurately estimated (Tymiński, 1997, p. 31). There are therefore significant discrepancies in the assessment of technical potential of renewable energy sources in Poland. Table 6.9 presents the technical potential for energy obtainable from renewable energy sources in Poland per year, as estimated by various institutions.

As can be seen from the analyses in Table 6.9, the technical resources of renewable energy in Poland are substantial. This generates the possibility to create in a relatively short period of time an additional stream of energy, which will supplement the national balance of electricity, heat and fuel production, and in the long term will become an alternative to conventional energy. The state of utilisation and the degree of interest in individual sources varies from one region of the country to another and is directly linked to the potential and local conditions for the development of a given area.

Notwithstanding development barriers, the share of renewable energy in Poland's gross final energy consumption is steadily increasing. The shares of renewable energy in final energy consumption in 2019 for the EU-28 countries and Poland were 10.2% and 11.6%, respectively. Between 2016 and 2019, there was an increase of 1.0 percentage point for the EU-28 countries and 3.5 percentage points for Poland (see Table 6.10).

The analysis of the capacity of RES installations in Poland clearly shows a steady, systematic growth. In 2016, the capacities of power stations using renewable energy sources in Poland amounted to 7902 MW, and in 2020, they amounted to 12,325 MW, showing an increase of almost 60%. In 2020, wind power capacity accounted for 51.1% of the total capacity of power stations using renewable energy sources. At the same time, hydropower installations enabled 7.9% and solid biofuels 6.0% of the renewable energy sources capacity. The installed capacity of solar power plants in 2020 accounted for 32.1% of the total renewable energy plant capacity and, compared

TABLE 6.9
The Technical Potential for Energy Obtainable from Renewable Energy Sources in Poland Per Year

Energy source	World Bank (1996)	BREC (2000)	KAPE (2007)
Unit	[PJ]	[PJ]	[PJ]
Biomass	810	895	530
Hydropower energy	30	43	30
Geothermal resources	approx. 200	200	170
Wind energy	4–5	36	250
Solar energy	370	1,340	170
Total	ca. 1,414	2,514	1,150

Source: Own elaboration adapted from Niemyski and Tatarewicz (2008, p. 56).

TABLE 6.10
Share of Energy from Renewable Sources in Total Final Energy Consumption (%)

Specification	2016	2017	2018	2019
Poland	8.1	7.8	11.7	11.6
EU-28	9.2	9.7	9.9	10.2

Source: Own elaboration adapted from Energy from renewable sources in 2020, 2021, p. 17.

TABLE 6.11

The Share of Renewable Energy Sources in the Total Renewable Energy Obtained in the Years 2016–2020

Specification	2016	2017	2018	2019	2020
			%		
Solar energy	0.7	0.7	0.7	1.1	2.0
Hydropower	2.0	2.4	1.4	1.4	1.5
Wind energy	11.8	13.9	9.1	10.6	10.9
Geothermal energy	0.2	0.2	0.2	0.2	0.2
Heat pumps	1.7	2.0	1.8	2.1	2.4
Biofuels:					
Solid biofuels	70.0	66.8	76.1	73.4	71.6
Liquid biofuels	10.1	9.9	7.5	8.0	7.8
Biogas	2.9	3.0	2.4	2.4	2.6
Municipal waste	0.7	1.0	0.8	0.8	1.1

Source: Own elaboration adapted from Energy from renewable sources in 2020, 2021, p. 36.

to 2016, increased more than 21 times and the electricity obtained from these plants almost 16 times (Energy from renewable sources in 2020, 2021, p. 55).

In both domestic RES generation and use, biomass is dominant. Among biomass, solid biofuels are key, accounting for 71.6% of RES energy harvested in 2016. Between 2016 and 2020, the share of wind energy decreased from 11.8% to 10.9%, biogas from 2.9% to 2.6%, and hydropower from 2.0% to 1.5%. At the same time, the share of solar energy increased from 0.7% to 2.0%. Table 6.11 presents data on the share of individual renewable energy carriers in the generation of energy from renewable sources between 2016 and 2020.

As can be seen from the analysis of the data in Table 6.11, in 2020, the largest increases in energy generation relative to 2016 were achieved by solar energy—an increase of 295% (i.e. by 7,769 TJ), municipal waste—an increase of 135% (i.e. by 3,452 TJ), heat pumps—an increase of 90% (i.e. by 5,912 TJ) and wind energy—an increase of 26% (i.e. by 11,565 TJ).

6.2.1 SOLAR ENERGY

Solar energy is a source of energy that is widely available. Throughout the country, there are opportunities to use solar energy for both domestic water heating and agricultural needs, as well as for the local production of electricity in photovoltaic cells.

The situation varies considerably across the country. The least favourable helioenergetic conditions are in the vicinity of Warsaw and the Upper Silesian region, while the most favourable conditions are in the seaside and Podlasie-Lublin area. The central part of Poland, that is, about 50% of the country's area, receives the irradiation of 1022–1048 kWh/m^2/year, while the southern, eastern and northern parts of Poland receive about 1000 kWh/m^2/year or less. Average annual values of sunlight in hours in selected Polish cities are presented in Table 6.12.

114 Management of Civic Energy and the Green Transformation

TABLE 6.12

Average Annual Values of Sunshine in Hours for Selected Cities of Poland

Meteorological stations	Sunshine in hours
Lublin	1929
Poznan	1875
Wrocław	1785
Warsaw	1693
Cracow	1583
Zakopane	1458

Source: Own elaboration adapted from Kuklo (2007, p. 19).

TABLE 6.13

Solar Energy—Photovoltaics

Specification	2016	2017	2018	2019	2020
Indigenous production (TJ)	446	596	1,082	2,558	7,048
Capacities of power stations using PV (MW).	187	287	562	1,539	3,955

Source: Own elaboration adapted from Energy from renewable sources in 2020, 2021, p. 55.

Obtaining and use of solar energy shows a steady increase. For example, in 2016, indigenous production in photovoltaics amounted to 446 TJ, while it reached 7,048 TJ in 2020 (an increase of up to 6,602 TJ). The installed capacity of solar power plants in 2020 represented 32.1% of the total generating capacity of RES power plants. Compared to 2016, it increased more than 21 times and the electricity obtained from these power plants increased almost 16 times (see Table 6.13).

6.2.2 HYDROPOWER

Most of Poland's hydropower resources are concentrated in the basins of Vistula (especially its right-bank tributaries) and Odra rivers. Table 6.14 presents the theoretical and technical potential of Polish rivers.

Between 2012 and 2020, hydropower generation fluctuated dynamically. This is a consequence of the instability of hydropower, which is dependent on the fluctuating water levels (see Table 6.15). For example, in 2012, hydropower production was 7,333 TJ, rising to 8,781 TJ in 2013, decreasing to 7,857 TJ in 2014, 6,596 TJ in 2015, 7,702 TJ in 2016, rising to 9,214 TJ in 2014 and remaining at around 7,000 TJ in the following years.

As of 31 March 2019, there were 765 renewable energy installations using hydropower with a total installed capacity of 981.5 MW. Most of these were plants with an installed capacity of less than 1 MW owned by private entrepreneurs, with only 47 owned by energy companies and water administrations. In turn, only 18 hydropower plants with a capacity of more than 1 MW are owned by private entrepreneurs, with 64 owned by energy companies and water administrations (Committee on Agriculture and Rural Development, no 335, 2019).

Renewable Energy in the Polish Energy Sector—Resources and Use

TABLE 6.14
Theoretical and Technical Potential of Polish Rivers

Specification	Potential (GWh/y)		Utilisation (in %)
	Theoretical	Technical	
Vistula river basin	16,457	9,270	56
Vistula	9,305	6,177	66
Left-bank tributaries	892	513	58
Right-bank tributaries	4,914	2,580	53
Other small rivers	1,346	–	–
Odra	2,802	1,273	45
Left-bank tributaries	1,615	619	38
Right-bank tributaries	1,540	507	33
Other	338	70	21
Rivers of the coastal zone	582	280	48

Source: Own elaboration adapted from Gołębiowski and Krzemień (1998).

TABLE 6.15
Hydropower (in TJ)

Specification	2012	2013	2014	2015	2016	2017	2018	2019	2020
Indigenous Production	7,333	8,781	7,857	6,596	7,702	9,214	7,092	7,050	7,626

Source: Own elaboration adapted from Energy from renewable sources in 2020, 2021, p. 78.

6.2.3 WIND ENERGY

Investment in wind power plants may also prove to be an interesting solution for the use of renewable energy. Conditions in Poland are relatively favourable. It is important to bear in mind that they are strongly linked to local climate and field conditions.

In terms of installed capacity, wind power plants lead the way. The installed capacity of wind farms in Poland amounted to 5.8 GW at the end of 2016. This represented more than 70% of the installed capacity of all RES for electricity generation. Between 2005 and 2016, wind-based energy was the fastest growing RES category in Poland—achieving a growth rate of nearly 70 times. Between 2016 and 2020, the dynamics of new investments in the wind energy sector decreased, and entities operating in the wind energy sector point towards the deteriorating situation of wind energy in Poland (The State of Wind Energy in Poland in 2016, The Polish Wind Energy Association, 2017, pp. 11–13). This is confirmed by statistical data, which indicate that wind power capacity remained at similar levels between 2016 and 2020 (see Table 6.16).

In 2020, wind power capacity accounted for more than 51% of the total renewable energy capacity.

6.2.4 GEOTHERMAL ENERGY

According to the concept of geothermal gradient, defined as the depth (measured in metres) at which the temperature increases by 10°C from the starting point, the

116 Management of Civic Energy and the Green Transformation

TABLE 6.16

Capacities of Power Stations Using Wind

Specification	2016	2017	2018	2019	2020
Indigenous Production (TJ)	45,315	53,673	46,076	54,384	56,880
Capacities of power stations using wind (MW)	5,747	5,759	5,766	5,838	6,298

Source: Own elaboration adapted from Energy from renewable sources in 2020, 2021, p. 55.

temperature varies according to the geological structure of the area, geotectonic development, chemical processes occurring deep in the Earth, proximity to volcanic phenomena and the presence of groundwaters. The average geothermal gradient in Poland is about 33 m/°C. As in other European countries, Poland has mainly sedimentary basins filled with geothermal waters with temperatures ranging from 20°C to about 90°C, with geothermal resources of approximately $2.9*1017$ J/km^2 (Basics of Geothermal Energy, 2021).

Geothermal energy in Poland in particular can be used for district heating. This offers great opportunities to reduce the consumption of traditional fuels, as well as to reduce emissions of pollutants, particularly low emissions. In addition, geothermal energy can be used extensively in therapeutics or recreation.

According to an analysis of statistical data, the extraction of geothermal energy in Poland is steadily increasing (see Table 6.17). In 2012, 661 TJ were extracted; in 2016, 930 TJ; and in 2020 already 1073 TJ. Geothermal energy was mainly used to meet heat demand. In 2020, 75% of geothermal energy was used in households, while 25% was used in trade and services.

In 2020, a relatively high position in obtaining of energy from renewable sources in Poland was achieved by heat pumps, ahead of solar energy, water energy or municipal waste energy. The use of energy extracted by heat pumps in 2020 was as much as 90.0% higher than in 2016 and amounted to 12,481 TJ (Energy from renewable sources in 2020, 2021, pp. 35–37).

6.2.5 BIOMASS

The technical potential of biomass available for energy purposes depends on the adopted economic model (including mainly the organisation of agriculture and forestry), the possibility to use increasingly efficient energy crop plantations and the availability of land that can be used for biomass energy production. Biomass in Poland is recognised as a renewable energy source with the largest resources and significant potential for development. Certainly, the primary direction of biomass use for energy purposes is and will be heat production. In the long term, solid biofuels are expected to be increasingly used also in cogeneration processes (Janowicz, 2006, p. 601):

Biomass can come from different sources, depending on the local resources: forests, agriculture or waste (Renewable Energy in Europe: Building Markets and Capacity. European Renewable Energy Council, 2004, p. 3). Under Polish conditions, the most relevant biomass sources include Janowicz (2006, p. 601):

- wood,
- crops for energy purposes,

Renewable Energy in the Polish Energy Sector—Resources and Use 117

TABLE 6.17

Geothermal Energy

Specification	2016	2017	2018	2019	2020
Indigenous Production	930	946	991	1,050	1,073
Commerce and Public Services	225	234	248	263	268
Households	705	712	743	788	805

Source: Own elaboration adapted from Energy from renewable sources in 2020, 2021, p. 46.

- oilseed crops,
- organic waste and residues.

The potential for energy agriculture is high. Agricultural biomass resources that can be used for energy purposes depend on cereal and rape crops. It is reported, for example, that between 10 and 14 t-ha-1 DM of straw can be harvested from 1 ha of various cereal crops. Average dry matter yields of hay from meadows are more than 12–15 t-ha-1, and under good conditions even more. Of the indigenous grasses, the common reed yields the highest, as its yields are valued at 12–30 t-ha-1. A hectare of maize, on the other hand, can produce 5,000 m³ biomethane per year (Janowicz, 2006, pp. 601–606).

At present, the area of perennial energy crops in Poland is estimated at about 4,000 ha, half of which are energy willow plantations. The potential supply of biomass from energy plantations is estimated at a level of about 50 million tonnes with an energy value of about 400 million GJ, which is equivalent to 20% of the coal currently used in the domestic energy sector. To obtain this amount of biomass, 1.3 to 1.5 million hectares of agricultural land would be utilised. The ever-increasing area of set-aside farmland and some extensively used grassland could serve as a base for the establishment of potential energy crop plantations. Also, the transitional climatic conditions in Poland with sufficient rainfall during the growing season are favourable for the cultivation of energy crops (Szczukowski and Tworkowski, 2006, p. 87).

As can be seen from the analyses of the statistical data from 2016 to 2020 (Table 6.18), the capacities of power stations using biomass were at similar levels.

Today's innovative waste management systems in highly developed countries are largely based on an integrated approach, where waste management is primarily waste incineration in either cogeneration or gasification facilities (Jarosiński et al., 2015, pp. 126–149). Unfortunately, Poland still remains behind among all European Union countries in terms of municipal waste management (Maśloch, 2013, pp. 11–28), where not only is most rubbish disposed in landfills but also one of the lowest levels of energy recovered from it is recorded among all EU countries.

According to an analysis of statistical data, the mass of municipal waste generated per capita in Poland is about 342 kg, and in 2020, 13.1 million tonnes of municipal waste were collected (Domańska et al., 2021, p. 4). Bearing in mind the experience of other countries, along with economic growth and the improvement of the standard of living, the production of municipal waste will also increase, which in highly developed countries annually reaches about 700–800 kg per capita.

TABLE 6.18

Capacities of Power Stations Using Biomass (MW)

Specification	2016	2017	2018	2019	2020
Solid biomass	727	709	735	732	734
Biogas	225	229	225	233	261
of which:					
Landfill gas	65	52	52	55	54
Sludge gas	77	71	72	74	86
Other biogas	83	106	102	104	121

Source: Own elaboration adapted from Energy from renewable sources in 2020, 2021, p. 55.

The municipal waste collected in 2020 was diverted to the following processes (Domańska et al., 2021, p. 4):

- recovery—7,732.8 thousand tonnes (59.0%), of which:
- recycling—3,498.6 thousand tonnes (26.7%),
- biological treatment processes—1,577.9 thousand tonnes (12.0%),
- thermal conversion with energy recovery—2,656.2 thousand tonnes (20.3%),
- disposal—5,384.1 thousand tonnes (41.0%), of which:
- by thermal conversion without energy recovery—166.4 thousand tonnes (1.3%),
- by storage—5,217.7 thousand tonnes (39.8%).

Huge amounts of municipal waste, by managing them for energy purposes, create completely new opportunities for both the Polish waste management and the energy sector. This is related, on the one hand, to the need to reduce the deposition of bio-degradable waste in landfills and, on the other hand, to the utilisation of the energy potential of such RES (see Maśloch, 2014, pp. 29–30). However, for this to happen, the right, satisfying conditions must be created for cooperation and the implementation of joint initiatives. As many stakeholders come into contact in the field of waste management: households, businesses or public sector entities, the key to success becomes the ability to compromise, cooperate and build network links based not only on commercial agreements or legal regulations but also on capital links or those related to pursuit of convergent goals. In addition, a number of technical problems concerning the possibility of using waste for energy purposes need to be solved.

With efficiency issues in mind, the Polish economy is entering a new stage of development in which new opportunities arise to break free from the inefficiencies of the energy sector. Analysing the structure of primary energy production in Poland over the past decades, one can see a systematic, albeit small, decrease in the role of hard coal and an increase in the share of oil, natural gas and renewable energy sources. The share of renewable energy sources in gross final energy consumption in 2020 in Poland amounted to 16.1% and increased by 7.43 percentage points compared to 2009. The

TABLE 6.19

Gross Final Energy Consumption from Renewable Sources in the Years 2016–2020 (in TJ and in %)

Specification	2016	2017	2018	2019	2020
Gross final consumption of RES for	232,653	236,860	367,959	360,528	356,160
heating and cooling	14.9%	14.8%	21.5%	22.0%	22.1%
Gross final consumption of	77,373	77,585	78,453	86,055	95,975
electricity from RES	13.3%	13.1%	13.0%	14.4%	16.2%
Gross final consumption of energy	22,545	28,876	41,969	46,740	47,202
from RES in transport	4.0%	4.2%	5.7%	6.2%	6.6
Gross total RES consumption	332,571	343,321	488,381	493,323	499,338
	11.4%	11.1%	14.9%	15.4%	16.1%

Source: Own elaboration adapted from Energy from renewable sources in 2020, 2021, pp. 57–58.

average annual growth rate of the share of energy from renewable sources in gross final energy consumption between 2009 and 2020 was 5.8% (see Table 6.19).

All of the aforementioned factors have a significant impact on the difficult situation in the Polish energy sector as a whole, which in the main bullet points can be characterised as:

- high dependence of the Polish energy sector on coal,
- low levels of trust in public authority,
- high energy intensity of the economy with low per capita consumption,
- high level of wear and tear of machinery and equipment,
- low efficiency of the power units,
- lack of alternative energy sources (including, inter alia, nuclear power plants),
- failure to ensure full energy security through dependence on oil and gas imports,
- small share of RES in the energy balance,
- a relatively underdeveloped civil society and thus civic energy.

6.3 LEGAL BASIS FOR THE USE OF RENEWABLE ENERGY SOURCES IN POLAND IN TERMS OF THE DEVELOPMENT OF CIVIC ENERGY

The development of civic energy, as a local energy community building its own energy supply systems and working on the basis of renewable energy sources, is linked to the establishment of appropriate legal regulations, both at national and international levels. Particularly in the case of Poland, a member of the EU, issues related to regulations and obligations concerning civic energy, renewable energy or energy transition are of particular importance. This is mainly determined by the policy of the community, which is becoming a global leader in climate change.

The EU is actively promoting Europe's transition to a low-carbon society and is updating legislation to facilitate the necessary private and public investment in the transition towards clean energy. An important issue for EU Member States is therefore that the transition to a low-carbon economy aims to create a sustainable energy sector that will stimulate economic growth, innovation and job creation. At the same time, it will improve quality of life, increase the offers available to consumers and enforcement of their rights and ultimately reduce energy bills.

When considering legal conditions for the creation of a civic energy sector, one should refer to regulations directly related to civic energy, but also take into account other areas of socio-economic life, including, among others, legal regulations in renewable energy, energy efficiency or environmental protection.

Considering the provisions characterising the EU's development to date in this area, it is noticeable that new elements have been included in the provisions of the current Fourth Energy Directive (FED). This directive forms a package titled 'Clean energy for all Europeans' (Clean energy for all Europeans. European Union, 2019). The directive significantly strengthens the position of the most vulnerable market participants (consumers) and extends their rights not only as energy consumers but also as entities taking an active part in the energy market (in particular in terms of self-generation and sale of electricity). One of the assumed solutions is the possibility to create and join 'citizen energy communities', so far unknown in EU legislation. Similar solutions are included in the provisions of the Renewable Energy Directive II (RED II), where the concept of 'energy communities' was introduced to define renewable energy entities.

The introduction of civic energy regulations or energy communities into the EU law is the result of key developments arising from:

- bringing the level of dispersed generation technology up to a point where it can meet the needs of energy consumers efficiently and cost-effectively,
- development of a civil society that is adequately educated, assertive and ready to take on new challenges and cooperate—including action to produce clean and green energy,
- 0 of dispersed renewable energy that requires harmonised rules of operation throughout the EU.

Citizen energy in Poland has found its place, inter alia, in the provisions of the Polish Energy Policy until 2040, where it is indicated that within the dispersed energy industry, two groups of active entities can be distinguished:

- active consumers (mainly individual entities, including, but not limited to, prosumers of RES, who generate energy for their own consumption, but have the possibility to return the surplus electricity generated to the grid or sell it, store energy and participate in other forms of activity. Active consumers form the backbone of civic energy. Increasing the number of renewable energy prosumers to 1 million in 2030 has been indicated as a target,

Renewable Energy in the Polish Energy Sector—Resources and Use

- energy communities—these are mainly collective entities, including, inter alia, energy clusters, energy cooperatives and other entities that organise themselves to generate electricity for their own use and undertake other activities (e.g. storage and energy sharing) for the benefit of members of their community. Increasing the number of such collective entities to 300 in 2030 has been identified as a target. (Energy Policy of Poland until 2040 (EPP2040), Ministry of Climate and Environment, p. 67).

Civic energy is therefore becoming an important element of the state's energy system, and its implementation can take place in Poland through the following available forms:

- prosumer and collective prosumer (legal basis: Act of 20 February 2015 on renewable energy sources),
- energy cooperatives (legal basis: Act of 20 February 2015 on renewable energy sources; Act of 16 September 1982 Cooperative Law),
- energy clusters (legal basis: Act of 20 February 2015 on renewable energy sources),
- virtual power plants (legal basis: Act of 10 April 1997 Energy Law),
- virtual prosumer (legal basis: Act of 20 February 2015 on renewable energy sources).

Of course, the catalogue given here has not been closed and further organisational and legal forms enabling the development of civic energy are expected in the near future (e.g. resulting from the RED II Directive—RES energy community).

7 Analysis of the Development of Civic Energy in Poland

CONTENTS

7.1 Introductory Remarks 123
7.2 Description of the Study 125
    7.2.1 Interviews with Representatives of Municipal and District Authorities 130
    7.2.2 Interviews with DSO Management 132
    7.2.3 Interviews with Residents/Entrepreneurs 133
7.3 Stakeholders Involved in the Implementation of Projects and the Potential Costs and Benefits of Participating in the System 133
    7.3.1 Society 134
    7.3.2 Enterprises 134
    7.3.3 Local Authorities 135
    7.3.4 State as a Regulator 137
7.4 Organisation of an Energy System Development Model Based on Local and Regional RES 138

7.1 INTRODUCTORY REMARKS

The following research workshop was used to investigate the potential and opportunities for the development of civic energy based on renewable and regional resources in Poland:

- **A systematic analysis of the literature on the subject,** which included the evaluation and interpretation of available published research relevant to the area covered in the book. This consisted mainly of studying publications available in libraries collections (the National Library in Warsaw, the University of Vienna, the University of Economics in Vienna, or the Library of Congress in Washington). Materials available in electronic full-text databases of scientific publications (e.g. Springer Link or Taylor & Francis Online) were also used. The nature of research objectives justified the need to use papers from different disciplines, that is, economics, finance and management, energy, environment or law.
- **Direct observation**, which the authors conducted as part of their work to prepare strategic management documents and investment projects in the field of dispersed energy or energy efficiency in Poland. The observation was conducted between 2008 and 2022, in municipalities and districts located

DOI: 10.1201/9781003370352-8

throughout Poland. It made it possible to collect the necessary information on the expectations and awareness of the public and private sectors, as well as the public in relation to civic or renewable energy. An important element of the study was the conducted and co-conducted meetings with entrepreneurs and inhabitants of individual regions of the country, regarding the possibilities of using renewable energy sources in civic energy.

- **Indirect observation**, which included material evidence and field visits. Material evidence consisted of information collected and descriptions of projects implemented, underway and under preparation by various entities in the field of civic energy. Field visits, on the other hand, were carried out to confront the material evidence with the facts.
- **Unstructured interviews.** The selection of people with whom unstructured interviews were conducted was purposeful and resulted from the need to recognise different expectations and points of view on the problems considered. In particular, interviews were conducted with representatives of local government units in Poland (municipal, county or provincial level), representatives of energy companies, companies using renewable energy sources and planning such investments, as well as meetings with representatives of the Ministry of Agriculture and Development or the Ministry of Climate and Environment).
- **Survey research.** A questionnaire survey was an important part of the study to gather information and data on the state of knowledge, interest and action taken regarding expectations of civic energy development opportunities. It was carried out nationwide. The survey was conducted using electronic tools from 1 June 2022 to 15 July 2022. The survey was conducted using a sample of 1,024 participants. Within the sample, 154 participants were self-employed. The sample included 606 urban residents and 418 rural residents.

The survey was conducted using the Association for Efficiency database and platform (www.stowarzyszenie-zmijewski.pl). The Association has been active in promoting the development of civic energy for many years and, as a pioneer in this field, has collaborated with a number of entities involved in a broadly defined energy transition.

- Given the number of respondents and their characteristics, the authors considered the aforementioned sample to be representative. Of course, the authors had in mind that in certain areas of the country, there may be specific social or economic conditions concerning civic energy, related, for example, to the dominant economic functions in some municipalities (e.g. mining functions). It is also worth noting that the respondents include individuals with knowledge of the basics of energy, individuals who are very aware of the role and importance of the energy transition.
- **Scientific reflection combined with induction and modelling processes,** which was used to develop a model for the development of civic energy in Poland and recommendations for the necessary measures for its implementation.

Analysis of the Development of Civic Energy in Poland

The research material used for collection and analysis resulted from the adopted aim of the study. The comprehensive and interdisciplinary nature of the study made it possible to collect the necessary information and data that made it possible to learn about the possibilities of development and expectations towards civic energy. It also contributed to the development of recommendations for actions that should be taken to develop an effective system for the use of civic energy in Poland.

7.2 DESCRIPTION OF THE STUDY

The purpose of the survey was to identify potential opportunities and threats to the development of civic energy in Poland and to identify priority strategic areas for civic energy development.

The survey consisted of ten problem questions. The individual questions were answered as follows:

> The first question: what forms of civic energy do you know, 989 respondents mentioned the individual prosumer, followed by the collective prosumer (635) and energy clusters (325). In contrast, only 125 respondents mentioned energy cooperatives and 132 respondents mentioned virtual power plants. This means, therefore, that as many as 90% of the respondents are not familiar with organised, and therefore more effective, forms of civic energy development.
> More than one answer could be selected for question 1. The distribution of individual responses is presented in Figure 7.1.

Respondents were also asked about the ownership of RES installations. There were 589 respondents (i.e. 57.5%) who own RES installations. The most popular were PV installations with 463 stakeholders (45% of the surveyed group) and solar panels, which were owned by 346 stakeholders (i.e. 33.8%). There were 435 respondents who did not own RES installations (see Table 7.1).

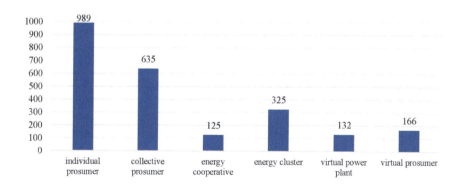

FIGURE 7.1 Responses to question 1: What forms of civic energy do you know? (More than one answer could be selected).

Source: Own elaboration.

126 Management of Civic Energy and the Green Transformation

TABLE 7.1

Responses to Question 2: Do You Have RES, If Yes Which Ones? (More Than One Source Could Be Selected)

Specification	Number of responses		Share of responses (in %)	
Answer I have RES/I do not have RES	Yes: 589	No: 435	Yes: 57.5	No: 42.5
Solar collectors	346		33.8	
PV	463		45.2	
Heat pump for central heating	185		18.1	
Heat pump for hot water	124		12.1	
Small wind power plant	2		0.2	
Small hydropower plant	3		0.3	
Biomass boiler	243		23.7	
I do not have RES	435		42.5	

Source: Own elaboration.

TABLE 7.2

Responses to Question 3: What Was (Will Be) Your Primary and Secondary Purpose for the RES Investment?

Specification	Main objective		Secondary objective	
	Number of responses	Share in %	Number of responses	Share in %
Ensuring energy security	0	0.0%	24	2.3%
Improving the quality of the environment	3	0.3%	285	27.8%
Reduction in operating costs	735	71.8%	134	13.1%
Profit-making	80	7.8%	35	3.4%
Legal requirements	47	4.6%	79	7.7%
Social pressure from the environment, fashion	0	0.0%	22	2.1%
Pressure from co-operators and contractors	45	4.4%	75	7.3%
Providing a higher quality of life	112	10.9%	186	18.2%
Other (what kind of?)	2	0.2%	32	3.1%

Source: Own elaboration.

The next question was devoted to the purpose that investors suggest (or will suggest) when undertaking actions in the field of renewable energy sources. According to the answers obtained (Table 7.2), the main motive is the desire to reduce operating costs (as many as 735 indications, i.e., 72%). More than 10% of respondents indicated, as the main motive for implementing RES investments, issues related to improving the comfort of living. Aspects related to improved quality of life resulting from RES were also frequently indicated as secondary objectives (186 indications, i.e., 18%).

In question 4, concerning means of financing the RES investment (respondents were asked to indicate the source of financing of more than 20% of the investment value and more than one source could be indicated), the majority of respondents confirmed that they financed or co-financed the investment with their own funds

Analysis of the Development of Civic Energy in Poland

(520, i.e., 88% among 589 RES owners). Bank and soft loans for the purchase of RES were also commonly used (198 indications, i.e., 34%), as well as EU grants (259 indications, i.e., 44%). It should be mentioned here that in the previous years, EU support measures have significantly changed the situation on the Polish renewable energy market, and in case of the civic energy sector, this was due to commonly implemented 'umbrella' projects, in which municipal governments acted as organisers and executors of initiatives enabling RES installations in individual residential buildings. Nowadays, more and more often, support instruments implemented with national funds (e.g. My Current initiative) support small-scale investments, as confirmed by the number of indications—185 or 31%. According to the survey, very rarely did respondents use loans and credits provided on a general basis (see Table 7.3 and Figure 7.2).

TABLE 7.3
Responses to Question 4: From Which Sources Was the RES Investment Financed (More Than 20% of the Investment—More Than One Source Could Be Indicated)

Specification	Number of responses	Share among those with RES (%)
Own resources	520	88.3%
Commercial credit	31	5.3%
Loan	25	4.2%
Bank loan, soft loan for the purchase of RES	198	33.6%
EU subsidy	259	44.0%
Domestic support resources (support programmes)	185	31.4%

Source: Own elaboration.

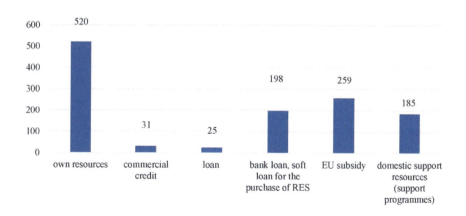

FIGURE 7.2 Responses to question 4: From which sources was the RES investment financed (more than 20% of investments—more than one source could be indicated).

Source: Own elaboration.

128 Management of Civic Energy and the Green Transformation

TABLE 7.4

Responses to Question 5: What Form of Citizen Energy Do You Represent?

Form of citizen energy	Number
Individual prosumer	580
Collective prosumer	0
Energy cooperative	0
Energy cluster	2
Virtual power plant	0
Virtual prosumer	0
Other	7
Total	**589**

Source: Own elaboration.

In response to question 5, regarding the form of civic energy represented, almost all RES users indicated individual prosumption. The answers obtained suggest that civic energy implemented in a corporate and organised manner is at a very early stage of development in Poland (see Table 7.4).

For question 6, concerning being a leader of change in RES technology implementation in the municipality/region, most respondents are convinced that it should be the municipality authorities (71% of indications). A much smaller group of respondents believes that it should be the entrepreneurs (17%) and only 5 respondents believe that the leader could be the inhabitants or an external coordinator (see Figure 7.3).

In the survey, respondents were also asked to answer a question on identifying the main barriers to the development of civic energy in Poland. For this question, the majority of respondents considered the lack of (insufficient) financial resources as the main reason. The lack of stability of legal regulations (240 responses) or the barrier of knowledge and skills (153 responses) were also considered relevant barriers (see Figure 7.4).

Particularly interesting answers were given to the question on what forms of public support are crucial for the development of civic energy. In this question, more than one answer could be indicated.

As can be seen from the answers provided, the expectations from the public sector in terms of civic energy development are to provide financial support on preferential terms, preferably with subsidies (891 votes). Also expected are initiatives to improve legal regulations (643 votes) and to take action to initiate new forms of civic energy organisation (536 votes) (see Figure 7.5).

The next question referred to the interest in developing civic energy. In total, 627 respondents, that is, 61%, have insufficient knowledge on the subject. Among the respondents, 372 (36%), express an interest and find individual prosumption most attractive (201 votes). In case of civic energy that can be implemented in an organised manner, the greatest interest is in energy cooperatives (90 indications) (see Table 7.5).

When asked further about the possibility of gathering trusted individuals to set up an organised civic energy structure, the majority, 77%, admitted that they do not have adequate conditions. Only 23% of respondents see such possibility (see Figure 7.6).

Analysis of the Development of Civic Energy in Poland

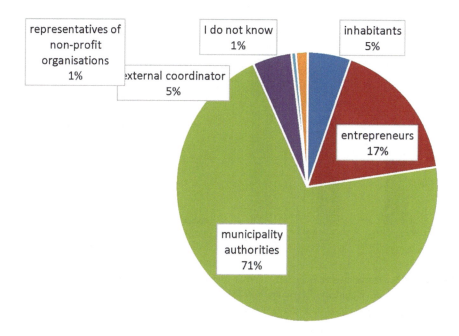

FIGURE 7.3 Responses to question 6: Who do you think should be the leader of change in the implementation of RES technologies in your municipality/region.

Source: Own elaboration.

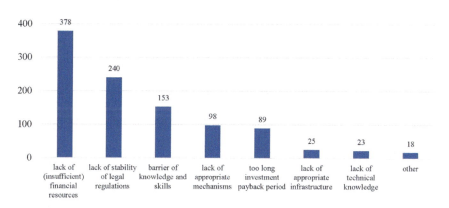

FIGURE 7.4 Responses to question 7: What are the main barriers to the development of civic energy in Poland?

Source: Own elaboration.

As part of the study, interviews were also conducted with both the authorities of the selected counties, as well as management of the DSOs (including PGE Distribution SA, Energa-Operator SA), residents and entrepreneurs from the selected municipalities. The interviews were conducted between 10 January 2018 and 30 August 2022.

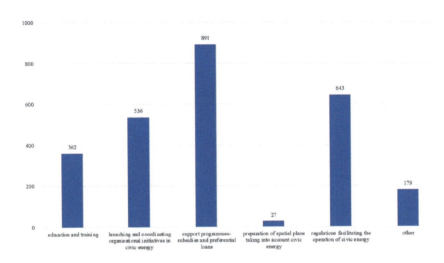

FIGURE 7.5 Responses to question 8: What forms of public sector support are key to the development of civic energy? (More than one answer could be indicated).

Source: Own elaboration.

TABLE 7.5
Responses to Question 9: Are You Interested in Further Developing Civic Energy—If So, in What Form?

Specification	Number of responses	Share in %
Not	25	2.5%
I don't know, I don't have sufficient knowledge	627	61.2%
Yes	372	36.3%
	Number of responses from the 'yes' group:	
Individual prosumer	201	
Collective prosumer	15	
Energy cooperative	90	
Energy cluster	22	
Virtual power plant	11	
Virtual prosumer	33	
Other	–	

Source: Own elaboration.

7.2.1 Interviews with Representatives of Municipal and District Authorities

The main objectives pursued by the municipal and district authorities are, on the one hand, to reduce the costs of maintaining municipal infrastructure and, on the other hand, to improve the quality of life in their area. Another important aspect of the measures taken is concern for the environment, in particular the state of ambient air. During working meetings with representatives of municipalities and districts,

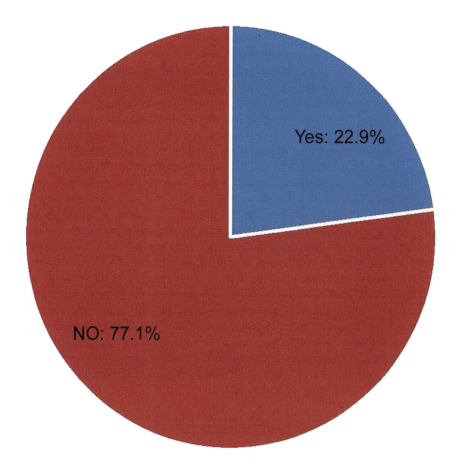

FIGURE 7.6 Responses to question 10: Do you have a group of trusted people (e.g. 10) among your immediate neighbours/co-workers with whom you would be able to make an effort to build organised civic energy structures (e.g. an energy cooperative).

Source: Own elaboration.

interest in the possibilities of developing renewable and civic energy in the area was assessed. This was also possible through an analysis of the basic documents prepared by the municipalities and districts, including, among others:

- local development strategies,
- strategies for solving social problems,
- revitalisation programmes,
- environmental programmes,
- low-carbon management plans.

The assessment of potential, preceded by a thorough analysis of the data and subjected to brainstorming, consequently led to the conclusion that many municipalities and districts have economic, social, environmental and infrastructural potential which entitles the local authorities to start working towards energy self-sufficiency in the future (see Maśloch, Maśloch, Kuźmiński, Wojtaszek, Miciuła, 2020).

It should be unequivocally emphasised that local government representatives advocate for an evolutionary approach to the creation of energy independence. Depending on the specific characteristics of the municipality or district, the energy needs that should be met by emerging RES energy initiatives have been prioritised differently. For some municipalities, the energy needs of public utility infrastructure are the priority, while for others the priority involves the needs of households.

A problem that local authorities see in the development of renewable energy in their area is lack of appropriate support schemes. The existing support programmes, which promote the cheapest solutions (e.g. biomass boilers or solar collectors) through cost indicators, are not as attractive from an economic and environmental point of view as, for example, integrated photovoltaic or heat pump systems. Furthermore, municipal and district authorities indicate that there is a lack of suitable advisory centres ready to provide specialised consulting services for innovative solutions in dispersed energy. According to the authors, in the statements of the representatives of local government units, the important role and importance of political conditions determining possible strategic, or investment initiatives was often clearly noticeable. Due to the fact that a significant proportion of local government representatives expect quick results and spectacular outcomes of the actions taken, despite seeing the undisputed advantages and necessity of RES investments in improving energy efficiency, the relatively long payback periods or time for positive effects to appear do not make this type of projects a priority.

7.2.2 Interviews with DSO Management

Meetings with representatives of the management boards of PGE Distribution SA and Energa-Operator SA confirmed the operators' readiness to cooperate with Energy Clusters and entities striving in other legal forms to achieve energy independence. A necessary condition for this cooperation is a partnership approach to joint initiatives. Representatives of DSOs shared their reflections on the fact that very often representatives of municipalities only raise their own arguments without seeing the problem on a wider scale (e.g. the indiscriminate use of LED lighting by some municipalities or ownership problems concerning power poles).

In the opinion of the DSOs, cooperation with individual entities or regions implementing renewable energy solutions requires an individual approach and consultation.

A starting point could be establishment of an agreement between a cluster, forming the initial organisational structure of areas where RES investments are made or initiatives are taken to achieve energy self-sufficiency, and the DSO.

The DSOs point to the belief commonly held by potential prosumers or other producers of energy from RES that the DSOs are reluctant to make new connections and cooperate with dispersed energy. Such approach, in the opinion of DSOs, is unjustified and may result to a large extent from a lack of knowledge or understanding of the functioning of the Polish energy sector. The lack of an integrated approach or lack of prosumer networking to enable effective management of generated energy creates

Analysis of the Development of Civic Energy in Poland 133

many problems for DSOs related to the appearance of 'wild energy', that cannot be effectively used at all times. Therefore, a way to significantly facilitate the operation of DSOs would be the emergence of entities that bring together and represent individual prosumers and RES producers in a way that enables efficient use of energy or provides reliable information on the planned amount of supplied energy.

7.2.3 Interviews with Residents/Entrepreneurs

The interviews conducted with residents and entrepreneurs, as well as the experience resulting from the author's extensive meetings about RES with residents and entrepreneurs in municipalities (e.g. municipalities such as Błonie, Serock, Mińsk Mazowiecki, Nadarzyn, Sulejówek, Sierpc, Nadarzyn, Jabłonna, Wojsławice and Baboszewo), prove a significant public interest in renewable energy. According to the interviews, residents as well as entrepreneurs expect clean, safe and cheap energy provided the renewable energy sector. A key problem slowing down the development of civic energy is the investment barrier. Therefore, many households make possible investments in renewable energy dependent on subsidy support. According to the declarations made, households or entrepreneurs accept an own contribution of up to 30–40% of investment costs. This is all the more important as, with appropriate organisation involving a move away from individual solutions towards comprehensive projects (e.g. construction of a photovoltaic farm or a biogas plant), these measures, thanks to the reduction of certain investment costs and the economies of scale, already make it possible to implement many investments without external support. In addition to the indisputable financial calculation, issues of comfort and quality of life remain particularly important for households. Particularly cumbersome and labour-intensive is the operation of coal-based boilers, which many households are willing to get rid of, even at the price of higher energy costs. For this reason, some apartment building owners are attempting to at least partially eliminate the inconvenience associated with the operation of coal-based boilers, for example by installing solar collectors or heat pumps for hot water.

For companies, in addition to the desire to reduce energy costs, the promotional aspect is also important. Companies using RES become more competitive and are seen as socially responsible. Another issue is that companies are increasingly faced with the need to switch to RES as a direct consequence of the requirements they face from other countries. This is a direct consequence of the demands placed on them by their counterparts in other countries, especially the Nordic countries, which increasingly make further cooperation conditional on at least several percentages of RES in the company's energy mix.

7.3 STAKEHOLDERS INVOLVED IN THE IMPLEMENTATION OF PROJECTS AND THE POTENTIAL COSTS AND BENEFITS OF PARTICIPATING IN THE SYSTEM

Effective civic energy development must be based on mutual cooperation and a network of economic and social links between the public, public administration and entrepreneurs. It should also be kept in mind that, regardless of the type of stakeholder

134 Management of Civic Energy and the Green Transformation

involved in the implementation of the project, each initiative is undertaken in a specific spatial location. It must, therefore, include opportunities for efficient and rational use of available human, capital, environmental and spatial conditions and resources.

7.3.1 SOCIETY

Social capital plays a key role in the development of civic energy. It is the degree of social acceptance and involvement, consisting, among other things, of the participation of the community in planning and implementation of individual strategies, plans and projects (starting with meetings, active forms of community engagement and including participation in investments), that determines the success of the implementation of civic energy model.

It is important to remember that households are not only consumers of energy but very often producers of energy as well. The development of renewable energy also creates new opportunities to involve communities in new development processes. Through the development of new economic forms, it also creates the possibility for residents to become involved in capital projects and in energy cooperatives or enterprises. This happens, for example, through acquisition of shares in companies involved in the production, sale and distribution of energy produced locally from renewable energy sources, or by taking their own initiatives to establish new business entities. In this respect, households can occupy an important place in the entrepreneurial structure of any local community. This approach creates new opportunities to increase energy efficiency. By participating in both the consumption and production of energy, the community can consciously make decisions that maximise the efficiency of its own resource management. In addition, important effects are undoubtedly achieved in terms of shaping social and entrepreneurial attitudes, improving the financial condition of households or engaging local communities in building local networks and, consequently, a civil society.

7.3.2 ENTERPRISES

Enterprises in the development of civic energy have the opportunity to play a variety of roles. On the one hand, they can be producers of energy used for their own consumption or share or sell their surplus energy. In this sense, whole companies, or their local subsidiaries/branches can be active participants in civic energy. They can also carry out activities for other entities in the civic energy sector.

A specialised activity of producing the energy for sale is becoming a completely separate aspect of the operation of enterprises. Another form of company activity can be manufacturing or providing services to the renewable energy sector. Activities can include the production of RES installations directly, as well as activities in sectors supporting renewable energy. Utilities in particular have a special role to play in the development of renewable energy, as they strive for energy self-sufficiency and can offer a completely new or higher standard of service to all customers they serve—including, of course, the residents.

The possibility of cooperation within the framework of public-private partnerships is of major importance for the development of civic energy. In this respect,

Analysis of the Development of Civic Energy in Poland

renewable energy offers entirely new opportunities for cooperation. The partnership can be either traditional in nature or take on various new, often 'soft' forms of cooperation, for example, within energy clusters.

An additional huge area of potential interest for enterprises can be in the field of education promotion. By running advertising campaigns, supporting education at all levels or carrying out charitable activities to protect the environment, the enterprises significantly shape social attitudes.

The quality of business management is also an important issue, especially with regard to the implementation of the concept of green management, which in practice may boil down to, among other things, conducting business processes with respect for sustainability aspects or optimising operations with a view of maximising the efficiency of the use of resources needed to achieve business objectives.

7.3.3 LOCAL AUTHORITIES

In the implementation of civic energy solutions, a special role is played by the public sector, especially local government units, which in Poland include the local (municipalities and districts) and regional (provinces) authorities.

This is because civic energy is created at the local level and that is where it functions and develops. Every inhabitant of a local authority (municipality or district) is an energy consumer. Each municipality, in turn, is obliged by law to supply energy to its residents (in places belonging to the municipality, e.g. schools and offices).

The local community should have the right to choose the source of the energy they purchase. Several solutions are available to them, with the easiest but capital-intensive one being a purchase of their own RES energy production equipment. Another option is to set up energy cooperatives. It is important for the local community to participate in the activities of their local authorities through bottom-up initiatives, encouraging local governments to take an interest in the energy transition. The local community should participate in creation of local laws and setting of locally relevant projects to support civic energy. Local authorities, on the other hand, need to manage energy consciously and prudently in their area, as well as cooperate with the large-scale energy sector.

Local government units (municipalities, districts or voivodeships), apart from the fact that they fulfil the tasks stipulated by the law, also have a real impact on the social and economic changes taking place in their area. They significantly influence spatial development and the state of the natural environment. Through the thoughtful creation of spatial plans, the development of documents related to strategic management, local energy planning and by acting in accordance with the will of its inhabitants, local government creates opportunities to solve the energy and environmental problems of its community.

The role of local authorities in the local social space naturally sanctions them to the role of a host of a given area, and the tools at their disposal provide real opportunities to successfully initiate new social and economic initiatives. Local government units should therefore become key initiators or co-participants in the development of new, network-based civic energy projects. This is determined by the fact that they can influence the development of civic energy through:

- rational strategic planning, including realistic strategies, programmes, plans and projects,
- creation of local laws to support a sustainable transformation based on civic energy,
- use of the widest possible range of instruments and tools to support and stimulate the activities of social and economic entities (especially in terms of infrastructure and environmental projects, energy, transport and communication),
- active participation of local authority subsidiaries, for example, municipal companies, in the development of energy based on renewable resources,
- active pro-environmental investment policy,
- organising and participating in educational and information campaigns that promote behaviour aimed at the development of civic energy, including environmental protection and air quality improvement,
- efficiency, commitment, competence development and cooperation of the administration at all levels in the implementation of the adopted programmes and plans.

The active role of local government, and in particular the charisma of local government leaders, becomes crucial when it comes to taking on new initiatives, including building new organisational structures. As suggested by the results of the survey discussed in this chapter, but also by meetings with local government authorities, a significant barrier to possible decisions is a lack of knowledge about the possibilities of building and the conditions that need to be met to join such initiatives. In addition, the fear of political risk associated with the possible failure of the project is not insignificant. Despite the substantial costs that a local authority has to bear and the barriers it has to overcome, active participation in the construction of local energy systems for each local authority brings a number of tangible benefits. These include opportunities to improve the quality of life in the area or to build a positive image of the public authority. It should also be noted that public security, including energy security, is the domain of the public sector. Therefore, the public has the right to expect sustainable access to secure and affordable energy.

Both energy prices and energy supply disruptions have a direct impact on the quality of life and the ability of the business sector to operate. Any price changes or energy supply disruptions significantly affect public confidence. A critical point in this respect becomes the threshold of public acceptability for spending on utilities, mainly energy. Exceeding the acceptability threshold means that consumers may stop paying for utilities, so that the whole system of financing them may collapse. Therefore, one of the key tasks of the public sector becomes the need to guarantee a secure energy supply for the public. In this respect, the role of the public sector, and the infrastructure under its responsibility, should be considered as an important element of local government participation in the creation of local energy systems based on renewable energy. The action of local government units within the emerging structure can also be linked to other municipal services. Municipal utilities can also be the leaders of the projects being built. As providers of services to the local and regional community, they have a direct impact on all residents of the region by ensuring their energy self-sufficiency (or striving towards energy self-sufficiency).

Analysis of the Development of Civic Energy in Poland

Savings made as a result of reduced energy expenditures can be reinvested in the development or improvement of municipal infrastructure. The creation of interconnected networks raises the standard of services offered while reducing the costs of these services.

The main motives influencing local government in deciding to co-create a local energy system based on renewable energy can include:

- the need to comply with legal requirements and environmental standards,
- rationalising the costs of maintaining public infrastructure,
- increasing the role of the public sector in the economy,
- social pressure,
- promotion of the region or local units,
- promotion of public officials in the community,
- energy security.

7.3.4 STATE AS A REGULATOR

Due to the degree of energy, environmental, fiscal or spatial regulation, there is a leading and crucial role for the public sector in the development of civic energy. In the energy sector, there is a direct and significant relationship between decisions made in public sector units and the private sector. It can therefore be clearly stated that the public sector, through regulation, has a decisive influence on the shape and possibilities of civic energy development.

In contrast, any costs of operating the energy system are not only borne by the private sector (directly or indirectly) but also have a key impact on the quality of life and competitiveness of businesses. They also condition the opportunities of future generations. The fundamental dilemma arising in this regard therefore concerns the extent to which both the needs and capitals at the disposal of society and economic operators are taken into account. They are not only consumers of energy but can also be energy producers, as well as providers and producers of services to the energy sector and energy-related projects.

However, it is important to distinguish between the role and approach to the central public sector and local government processes. In the case of the state, the focus is mainly on the problems of organising civic energy from the legal point of view and conditioning its cooperation with conventional energy.

Due to the degree of difficulty of the undertaking to create an innovative organisational structure, it is necessary to prepare, at a central level (e.g. Ministry of Agriculture and Rural Development in case of energy cooperatives), guidelines and model solutions for the most relevant aspects of the issue in question. A key tool, therefore, becomes the creation of a database of good practices and a platform for the exchange of information, promoting the initiatives undertaken and enabling consultation and joint problem-solving. An important role in the effective development of civic energy in Poland may also be played by the introduction of a dedicated tariff for energy self-sufficient regions (or those striving towards energy self-sufficiency), which will cooperate with conventional energy only under specific conditions and rules. In addition, the following issues also seem important:

138 Management of Civic Energy and the Green Transformation

- development of spatial planning standards, taking into account local and regional civic energy potential,
- amendment of the construction law, along with determination of the justification for the extension of the network infrastructure,
- targeting subsidy support schemes for projects that comprehensively address the problems of civic energy investment in a given area and making their award conditional on the socio-economic and environmental outcomes achieved,
- the development of fiscal incentive instruments for those investing in renewable energy sources as part of civic energy.

7.4 ORGANISATION OF AN ENERGY SYSTEM DEVELOPMENT MODEL BASED ON LOCAL AND REGIONAL RES

The primary objective of creating a civic energy structure in economic practice is to move towards energy self-sufficiency in a sustainable manner. In order to achieve this goal, it becomes necessary to cooperate and develop new relationships between all stakeholders in order to overcome the limitations in accessing energy (especially regionally available and environmentally friendly energy). As can be seen from the results of the surveys and interviews presented in this chapter, many developmental barriers can be identified, the overcoming of which will be necessary for civic energy in Poland to develop effectively. These include problems related to:

- limited knowledge of the genuine potential and opportunities for civic energy development,
- the lack of adequate human resources prepared to work for renewable energy, including leaders making the effort to implement effective solutions,
- education barrier and the associated information barrier on renewable energy,
- the availability of installations on the market (particularly in the context of public procurement),
- a barrier to investment costs. (In most cases, capital expenditure is very high relative to the capacity of potential investors. Although operating costs are low, the need for significant capital expenditure is a significant barrier to investment),
- legal regulations (unstable legislation in the field of renewable energy and related areas—e.g., spatial planning or agriculture and forestry),
- low levels of public confidence.

When embarking on the construction of organisational models for civic energy, it is useful to recognise certain characteristics that represent the expectations and capabilities of different stakeholders:

I. Objectives of the entities in the system

Stakeholders engaging directly or indirectly in the process of creating an energy structure based on civic energy may be driven by different objectives. These objectives may be both the result of the area of interest of individual stakeholders, as well

Analysis of the Development of Civic Energy in Poland

as the result of individual aspirations of the persons taking action. The main objectives can include the following issues:

- economic (improving financial condition, increasing competitiveness, adapting to the requirements of co-operators),
- social (e.g. improving service delivery, improving quality of life, tackling energy poverty),
- environmental (e.g. improving ambient air, the quality of water or soil),
- political (e.g. implementation of economic strategies, developing a position in the society).

It should also be kept in mind that the development of civic energy may also hit the current interests of many groups, including but not limited to, the conventional energy sector. In this regard, it is important to identify and counteract initiatives intended to limit the opportunities for civic energy development.

II.   Resources and competences

Changes taking place in the energy system should be evolutionary, not revolutionary. Such approach requires the construction of appropriate development strategies.

In the case of organised structures, for the sake of transparency of operations, entities operating within the created structure should strive to create a framework of specific corporate governance, acceptable to all and guaranteeing transparency of rules and procedures in the created structures.

To this end, the civic energy development process must take into account:

- environmental resources and conditions,
- human capital,
- energy technologies emerging within the reach of those concerned,
- energy infrastructure available in the area,
- investment opportunities and capital accumulation skills,
- the ability of individual stakeholders to take advantage of development opportunities and possibilities,
- public sector policy.

III.   Key actions and initiatives

In the area of interest of civic energy entities operating based on renewable energy sources, apart from direct investments in renewable energy, activities increasing energy efficiency as well as promotional and educational activities are also important. A specific type of activity organised within the established structure may be that of providing services and manufacturing for the civic energy technology and related industries.

IV.   Benefits achieved

Participation in and cooperation with civic energy can generate various benefits, which, due to their specificity, should be categorised according to social, economic, financial, political or environmental impacts. An important issue in the process of civic energy development is adoption of a principle that benefits cannot be accumulated by a narrow group of entities, and that all stakeholders operating in a given area should be beneficiaries.

V. Period of engagement in the venture

Building an effective civic development business model using local and regional resources is a long-term process. Depending on the individual preferences of different entities, the entry and exit barriers should not be excessive, which can provide an opportunity for different entities to gradually become involved in ventures, depending on their preferences, resources or motivations

VI. Social architecture of the venture

In the created civic energy system, it becomes important to be able to build new organisational structures that will provide adequate conditions for the functioning of entities cooperating for the development cooperation. In this aspect, it will be important to develop appropriate management systems and select adequate human resources, as well as the ability to motivate and develop.

In an effective model for RES-based civic energy development, stakeholders form a mutual network of capital, interpersonal or organisational links, which can be reflected in the ARA model (Håkansson and Snehota, 1995, pp. 43–44) (see Figure 7.7).

The network arrangement presented in Figure 7.7 consists of both the content and functions of the relationships. The content of the relationship includes connections in terms of activities, entities and resources. The functions of the relationship, on the other hand, focus on aspects concerning the relationships between different stakeholders, connections between them and impact of the network connections involving all network members. The role of the network is to integrate the key areas

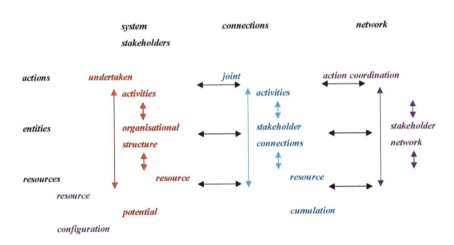

FIGURE 7.7 Analysis of connections in the local civic energy structure presented using the ARA model as an example.

Source: Own elaboration adapted from Håkansson and Snehota (1995, pp. 43–44).

of competence of individual members and to create structures for the intra-network diffusion of resources in order to achieve the synergy effect.

The activities of stakeholders in civic energy are interconnected and interdependent. Therefore, individual actions taken by individual entities are not fully effective. Maximising the effects can only be achieved through joint action by as large group of entities as possible. They should therefore be coordinated and managed accordingly. At a later stage of civic energy development, the coordination and management should be increasingly improved, in terms of both investment and management of the resulting energy infrastructure and energy production. Structure of the network is built by consciously interdependent entities. However, the network is neither a loose bundle of relationships nor a hierarchical structure of a dominant nature. It can be both a regulatory and an organisational structure. Within its sphere of influence, it takes initiatives to represent interests of the stakeholders in the structure. These activities can be undertaken using both ownership relationships and voluntary cooperation agreements. The cementing point of organisation of a civic energy network structure in a local or regional economy should be the common goal of working towards energy self-sufficiency. To this end, it is necessary to create bodies that bring together civic energy entities—for example, in chambers of commerce, associations or unions.

The nature of civic energy based on renewable energy determines the need for cooperation between entities in both the public and private spheres. In addition to developing a platform for cooperation between public sector entities and businesses, a positive political and social climate for change is becoming a key to success. Completely new opportunities are also opening up for direct capital involvement of citizens in development processes, which, with the support of research and development units, universities and chambers of commerce bringing together entities interested in implementing solutions in civic energy, have the chance to create a new quality in the process of energy generation, distribution and consumption.

The construction of any system should start by defining the motivations that led to the decision to take the appropriate action. Entities initiating change may be guided by different motivations for action. Motivations translate into choices and very often determine the direction of changes taking place. In the construction of energy systems, motivations can range from the very individual and down-to-earth, such as the desire for profit, to social (e.g. to improve air quality or quality of life), environmental (e.g. care for the natural environment) or patriotic motives (related to ensuring energy security that is key for national security). Decisions to make changes may also be dictated by the available conditions for access to natural resources that have a key impact on RES potential at a given time and place. An adequate analysis and assessment of the potential has a key impact on the whole process of building a business model.

The motivation of those initiating the construction of systems or preparing to make investments in civic energy clearly translates into the adopted objectives of action. The objectives adopted, on the other hand, determine the nature of changes. Depending on the actions taken, these changes may involve replacement, modernisation or development investments.

The next procedural step is the decision-making stage concerning destination of the energy produced. Depending on the purpose for which the energy will be

142   Management of Civic Energy and the Green Transformation

produced, choices are made regarding use of appropriate technologies, as well as formal and organisational measures to be taken. The most common energy destination decisions include:

- decisions of a private nature to produce energy solely for own needs and use,
- private-commercial decisions, where energy will be produced both for own use and the use of others without sale, as well as with a view of sale.
- commercial decisions to generate energy for profit with a view of sale.

These decisions condition functioning of the energy system, which in theory can operate in closed and open systems. At present, however, it is not possible in economic practice for completely closed efficient hermetic systems to operate, in which all the components and services required for operation will come from a small local or regional community.

The directions and uses of energy correspond directly to the benefits that entities intend to obtain from the projects. At the same time, the main benefits that can be achieved from investments in civic energy are:

- financial benefits from energy sales,
- reducing the cost of purchasing energy from conventional energy sources,
- avoiding penalties or difficulties associated with failure to meet environmental standards,
- environmental benefits,
- bridging the 'gap' in conventional power generation,
- stabilisation of the energy system (with the development of appropriate energy storage technologies),
- improving the innovativeness and potential of the economic entity,
- improving the image of the surrounding area,
- improving spatial aesthetics,
- increasing the value of assets,
- ensuring energy security,
- ensuring energy self-sufficiency,
- compliance with legal or formal requirements.

This stage is followed by the need to make a key decision on how to manage the infrastructure and energy. As a rule, most entities in Poland decide to implement and run the project by themselves and on their own account. Obviously, such decision has a number of advantages and does not raise concerns about the need for cooperation or dependence on the will and decisions of other entities. However, for obvious reasons, acting alone does not offer the possibility of gaining economies of scale in implementation or of using the knowledge, experience and also the cooperation capacity of the entities. Very often cooperation is unavoidable (e.g. cooperation with DSOs). Therefore, a question arises whether to establish such cooperation on an ad hoc basis or to make the effort to build advanced organisational or capital structures. In view of the above, the key aspect becomes the ability to cooperate and decide to operate within a specific formal and legal structure. These activities can start with the

Analysis of the Development of Civic Energy in Poland 143

establishment of a cluster, bringing together the first entities declaring a joint effort to invest in RES. The cluster is a voluntary agreement to analyse the goals, potentials and resources of individual members and their ability and willingness to engage in joint development activities.

Energy cooperatives, or other legal forms, also offer particular opportunities for action to achieve the desired goals in an organised and efficient manner.

Building local and regional system solutions for civic energy will also require the involvement of individual entities in promotional and educational activities. Activities in this area should be carried out in a very broad manner. On the one hand, lobbying for actions and initiatives, and on the other, comprehensive education on the principles of efficient energy production and consumption. This activity should take place both through cooperation and exchange of good practices with other regions operating in local and regional energy structures or cooperation with universities, institutes, chambers and organisations bringing together entities working for the development of RES. There is, of course, scope for cooperation with self-producers, but given the scale and degree of difficulty, this will be ineffective and inefficient.

An important aspect of networking activities for the pursuit of energy independence using local and regional civic energy resources and building competitive advantages for communities is the ability to create further economic ventures within the emerging systems. In this respect, the problem must also be considered in very broad terms. The fundamental issue remains the legal form of the new economic forms on the one hand, and the scope of activity and range of operations on the other. At the same time, the issues to be analysed must be considered together, and the choice of the appropriate legal form must depend on the scope and scale of activity of the entity in question. This approach to the development of efficient in terms of energy and socio-economic aspects local and regional renewable energy systems makes it possible to create businesses that will:

- produce and provide services to their own energy system,
- produce and provide services to their own energy system and others operating within and outside the system's influence,
- carry out business ventures outside the system.

How to cooperate with the DSO also becomes a key issue. In this respect, entities may adopt solutions that are independent or dependent on the DSO. The first type is largely complementary to the activities of the DSO and consists of cooperation involving use of its services. In case of large organisational structures, renewable energy, however, cannot at present reliably secure energy (especially electricity) demand throughout the year. It will therefore be necessary in the coming decades to maintain conventional power plants as stable energy sources. This therefore requires the development of effective solutions for the cooperation of dispersed energy with conventional energy and DSOs. For this purpose, it becomes necessary, among other things, to introduce new energy tariffs dedicated to RES-based energy structures, in which conventional energy will be treated as a supplementary or system-stabilising element, guaranteeing energy security.

A further challenge for organised local or regional civic energy structures may be to take over the distribution network or, in the extreme case, to build their own. Ultimately, any region that wishes to remain energy self-sufficient will have to cope with this task.

The problem of implementing RES investments is also becoming fundamental. Due to the necessary competence and equipment, it is in practice very difficult and risky to carry out investments on the individual entities' own account. The commissioned system therefore remains. However, observing the installation market and the investment pressures not only in the RES sector but in the entire construction industry, there is a problem with securing appropriately prepared teams to guarantee high quality installation services. Economic practice to date indicates that in many cases it is not possible to invite suitable contractors to investments, and potential applicants do not agree to the conditions offered and, taking advantage of the lack of competition, submit bids that significantly exceed cost estimates.

When it comes to new RES investments, access to investment funding sources is crucial. The capital that can be engaged in building RES energy systems in Poland requires a rational and efficient approach where both environmental, social and economic aspects are maximised. Such approach largely excludes the individual approach in favour of team initiatives, organised in a long-term and sustainable manner.

Conclusion

The dynamic political, demographic, social and economic changes taking place around the world are causing an increase in demand for energy, which in turn contributes to the need to seek out new possibilities for obtaining and using it. Thus, the paradigm of modern energy systems is changing, with an increasing tendency to abandon the hierarchical structure of the energy industry in favour of solutions of a dispersed nature, using endogenous resources, owned and managed by local communities.

In this sense, civic energy, using local, small-scale installations to generate electricity or heat from renewable sources, is becoming particularly important. Civic energy also enables the participation of local communities in larger RES projects, where local governments or other stakeholders come together to operate in the market in an organised manner (e.g. within energy cooperatives). The advantages of civic energy include equal access to energy for all, improved energy efficiency for households and SMEs, and a positive impact on the environment and quality of life in the area.

In order to use the existing technical potential of renewable energy sources for the benefit of civic energy, it is necessary to ensure appropriate conditions favourable to their development, to increase financial effort for research and development of technologies and to create a system of effective support for investment and organisational undertakings, supporting the development of various forms of civic energy.

Making a change in the energy sector towards sustainability is not easy. It is a major challenge, which is a complex and long-term process that requires financial and organisational effort made by public authorities, companies and societies as a whole. In order to make the right choice between alternative energy technologies, it becomes essential to take all costs into account. This necessity arises from the increasingly important economic rationale in the strategic decisions of individual countries, whose role, given the importance of energy for socio-economic development, continues to grow. As a result of the energy transition, political considerations are becoming less important and are part of a global trend whereby the transition is taking place from a monopoly energy sector to a competitive market in which civic energy, mainly based on renewable energy sources, is playing an increasingly important role.

The problem of energy transition and the need to develop civic energy is particularly relevant in Poland. The literature reviews and research carried out, as well as creative thinking, made it possible to achieve the aim of the work and indicated the necessity of launching new mechanisms in Poland, which will allow for a real reduction in primary energy obtained from non-renewable sources in favour of an increase in the level of production and consumption of energy coming from renewable sources and creation of conditions in which civic energy will be able to effectively join the development of the national energy system.

DOI: 10.1201/9781003370352-9

The considerations and analyses carried out in this chapter show that in Poland, under the current socio-economic conditions, it is possible to implement a process of reconstruction of the Polish energy sector in line with changing paradigm in the energy sector. However, it should be remembered that the structure of energy production and consumption in Poland is different from that of highly developed countries. This is mainly due to the dominant share of coal and lack of nuclear power. This state of affairs additionally hinders and exacerbates the situation of the country, which in the coming years will have to make a civilisational leap, expressed in the necessity to rebuild the entire energy sector, increasingly opening up to RES.

The process of rebuilding the Polish energy sector must be carried out in accordance with energy and public security policies. Civic energy, due to its dispersed, local character, can already be an important factor positively influencing energy security and, in connection with the innovative character of the sector, can also contribute to building competitive advantages of the Polish economy.

Renewable energy generation in civic energy structures is becoming increasingly competitive with conventional or nuclear power. The changes taking place in this area are taking place despite the presence of many barriers. One factor limiting the development of civic energy is the associated relatively high level of initial investment effort, which encourages small investors (potential prosumers and cooperatives) with limited financial resources to choose non-renewable solutions. However, unlike conventional technologies, the outlays required to invest in wind, heat pumps, photovoltaics or certain biomass-based technologies are steadily decreasing. This is determined by the phenomenon of economies of scale, increased technological efficiency and the increasing level of knowledge of civic energy solutions in investors.

Civic energy therefore has a significant impact on the standard and quality of life of residents and meeting social needs. This impact is particularly noticeable with dispersed development in rural areas. As the study showed, the public is interested in using RES, both in their own households and in their businesses. At the same time, individual entities may have different preferences in terms of the choice of source and method of energy production and supply. In addition to the increasingly advantageous desire to achieve energy self-sufficiency, society is also paying attention to the impact of the energy sources used on human health and the environment, as well as the quality of life they provide. The desire to improve the condition of budgets, for which the prospect of being a prosumer is becoming increasingly more attractive, is also becoming significant. In this respect, solutions allowing investors and prosumers to organise themselves into networks, such as purchasing groups, energy clusters, energy cooperatives and other forms that maximise the benefits of scale, also offer completely new opportunities. A key problem that needs to be seriously addressed is Poland's relatively underdeveloped civil society and its high degree of distrust of the possibility of acting in an organised way, whether with private individuals, public authorities or businesses.

The lack of innovative organisational solutions in the energy sector and the further development of energy operating models using existing organisational, technical or ownership experience may lead to a collapse in both the energy sector and the economy as a whole, which will not be able to supply energy for an innovative economy and an increasingly mobile and technology-enabled society. In addition, for many

Conclusion 147

businesses, not having access to 'clean energy' will be a disincentive to engage with those expecting their partners to use RES or zero-carbon energy. Benefits from the introduction of green technologies will therefore be naturally implied by the extent to which renewable energy is used. This also creates new opportunities for individual regions which, with renewable energy, can become a business partner or an interesting location for investment.

The rapid development of civic energy is already posing challenges to the energy system in particular, which has to adapt to increasingly decentralised and diversified production. The development of RES makes an important contribution to improving the energy infrastructure and to significantly increasing competitiveness in conventional power generation. The conventional energy industry, when competing with renewable energy producers, is increasingly forced to change its attitude towards its customers or the way it produces and provides services. This means that the decisive factor in purchasing energy is no longer its cost, but such issues as impact of the energy supplied on the environment, health, quality of life, or, finally, development opportunities of the local community.

Technologies that are already available today make it possible to build financially, socially and environmentally efficient energy mixes. Investment in civic energy is also becoming an essential element in increasing the innovation and competitiveness of any economy, and undertaking and supporting development activities in this area is necessary because of:

- securing growing energy needs,
- the need to improve energy security through diversification of energy sources and at least partial independence from imports,
- improving the environment,
- the need to meet the requirements and obligations of international agreements and commitments adopted within the framework of international organisations,
- the development of an innovative, knowledge-based economy, using state-of-the-art technologies and with skilled staff working in the energy sector,
- the continued rise in the standard of living of societies and the need to tackle energy poverty.

Bibliography

Activity Report. The President of the Energy Regulatory Office in Poland, Warsaw, 2017.

Activity Report. The President of the Energy Regulatory Office in Poland, Warsaw, 2022.

Albadi M.H., El-Saadany E.F., A summary of demand response in electricity markets. Electric Power Systems Research, Vol. 78, no. 11, 2008.

Alotto P., Guarnieri M., Moro F., Redox flow batteries for the storage of renewable energy: A review. Renewable and Sustainable Energy Reviews, Vol. 29, 2014.

Alvarado R., Deng Q., Tillaguango B., Méndez P., Bravo D., Chamba J., Alvarado-Lopez M., Ahmad M., Do economic development and human capital decrease non-renewable energy consumption? Evidence for OECD countries. Energy, Vol. 215, Article ID: 119147, 2021.

An Integrated Industrial Policy for the Globalisation Era, Communication from the Commission to the European Parliament, The Council, The European Economic and Social Committee and the Committee of the Regions, COM/2010/0614 final, 2010.

Andreae M.O., Biomass burning – Its history, use, and distribution and its impact on environmental quality and global climate. [In:] J.S. Levine (ed.). Global Biomass Burning: Atmospheric. Climatic and Biospheric Implications. MIT Press, Cambridge, 1991.

Art. 1, Dyrektywa 2006/32/WE Parlamentu Europejskiego i Rady z dnia 5 kwietnia 2006 r. w sprawie efektywności końcowego wykorzystania energii i usług energetycznych oraz uchylająca dyrektywę Rady 93/76/EWG, 2006.

Art. 2, Dyrektywa Parlamentu Europejskiego i Rady 2009/28/WE z dnia 23 kwietnia 2009 r. w sprawie promowania stosowania energii ze źródeł odnawialnych zmieniająca i w następstwie uchylająca dyrektywy 2001/77/WE oraz 2003/30/WE, 2009.

Badouard T., Oliveira D.M., Yearwood J., Torres P., Altmann M., Cost of Energy (LCOE), Publications Office of the European Union, Final Report, Luxembourg, 2020.

Bahn O., Barreto L., Kypreos S., Modelling and assessing inter-regional trade of CO2 emission reduction units. Environmental Modeling & Assessment, Vol. 6, no. 3, 2001.

Balcer K., Gajowiecki J., Klera-Nowopolska M., Magiera E., Tyszkiewicz K., Skolimowsk M., The State of Wind Energy in Poland in 2016. The Polish Wind Energy Association, Warsaw, 2017.

Banks F.E., Energy and Economic Theory. World Scientific, London, 2015.

Barczak B., Koncepcja oceny efektywności struktur sieciowych. Wydawnictwo Uniwersytetu Ekonomicznego w Krakowie, Kraków, 2016.

Bartkowiak G., Psychologia zarządzania. Wydawnictwo Akademii Ekonomicznej w Poznaniu, Poznań, 2009.

Bartkowiak R., Długie cykle Kondratiewa, Studia i Prace Kolegium Zarządzania i Finansów. Szkoła Główna Handlowa, no. 17, 2000.

Basics of Geothermal Energy, 2021, www.energia-geotermalna.org.pl, accessed: 05.02.2021.

Bauer B., Fischer-Bogason R., Voluntary Agreements and Environmental Labelling in the Nordic Countries. Tema Nord, Kobenhavn, 2001.

Bekkeheien M., Håland Ø., Klovenings R., Energy demand patterns towards 2050. [In:] Energy. The Next Fifty Years. OECD, Paris, 1999.

Berent-Kowalska G., Jurgaś A., Kacprowska J., Szymańska M., Moskal I., Energy from Renewable Sources in 2020. GUS, Warsaw, 2021.

Berent-Kowalska G., Kacprowska J., Kacperczyk G., Jurgaś A., Energy from Renewable Sources in 2008. GUS, Warsaw, 2009.

Bergmann A., Public Sector Financial Management. Pearson Education, Essex, 2009.

Bergstrom J.C., Randall A., Resource Economics: An Economic Approach to Natural Resource and Environmental Policy. Edward Elgar Publishing, Cheltenham, 2016.

Bertoldi P., Boza-Kiss B., Analysis of barriers and drivers for the development of the ESCO markets in Europe. Energy Policy, Vol. 107, 2017.

Bianco V., Manca O., Nardini S., Electricity consumption forecasting in Italy using linear regression models. Energy, Vol. 34, no. 9, 2009.

Bielawa A., Przegląd kryteriów i mierników efektywnościowych przedsiębiorstw nastawionych projakościowo. Studia i Prace Wydziału Nauk Ekonomicznych i Zarządzania, no. 34, 2013.

Bielecki S., Skoczkowski T., Europejskie projekty rozwoju inteligentnych sieci energetycznych. Obraz ogólny i miejsce Polski. Polityka Energetyczna, Vol. 17, no. 4, 2014.

Bigliardi B., Dormio A.I., An empirical investigation of innovation determinants in food machinery enterprises. European Journal of Innovation Management, Vol. 12, no. 2, 2009.

Bilgili M., Ozbek A., Sahin B., Kahraman A., An overview of renewable electric power capacity and progress in new technologies in the world. Renewable and Sustainable Energy Reviews, Vol. 49, 2015.

Blaug M., Teoria ekonomii. Ujęcie retrospektywne. Wydawnictwo Naukowe PWN, Warszawa, 2000.

BloombergNEF, https://about.bnef.com, accessed: 27.09.2022.

Boisvert R.N., Cappers R., Goldman C. et al., Customer response to RTP in competitive markets: A study of niagara mohawk's standard offer tariff. The Energy Journal, Vol. 28, no. 1, 2007.

Bolinger M., Wiser R., O'Shaughnessy E., Levelized cost-based learning analysis of utility-scale wind and solar in the United States. iScience, Vol. 25, no. 6, 2022.

BP Energy, BP Statistical Review of World Energy, BP Energy, 2021, https://www.bp.com/, accessed: 27.09.2022.

BP Energy—Outlook, 2017 edition, www.bp.com, accessed: 27.09.2022.

Brunekreef B., Holgate S.T., Air pollution and health. The Lancet, Vol. 360, no. 9341, 2002.

Bruvoll A., Medin H,. Factors behind the environmental Kuznets curve: A decomposition of the changes in air pollution. Environmental and Resource Economics, Vol. 24, no. 1, 2003.

Brzóska J., Krannich M., Modele biznesu innowacyjnej energetyki. Studia Ekonomiczne. Zeszyty Naukowe Uniwersytetu Ekonomicznego w Katowicach, no. 280, 2016.

Brzozowska A., Havrysh V., Kalinichenko A., Stebila J., Agricultural residue management for sustainable power generation. The Poland Case Study Applied Sciences-Basel, Vol. 11, no. 13, 2021a, www.mdpi.com.

Brzozowska A., Havrysh V., Kalinichenko A., Stebila J., Life cycle energy consumption and carbon dioxide emissions of agricultural residue feedstock for bioenergy. Applied Sciences-Basel, Vol. 11, no. 5, 2021b, www.mdpi.com.

Brzozowska A., Korczak J., Kalinichenko A., Bubel D., Sukiennik K., Sikora D., Stebila J., Analysis of Pollutant Emissions on City Arteries—Aspects of Transport Management. Energies, Vol. 14, no. 11, 2021c, www.mdpi.com.

Burchard-Dziubińska M., Ekologiczne determinanty rozwoju gospodarczego. [In:] R. Piasecki (ed.). Ekonomia rozwoju. PWE, Warszawa, 2007.

Burger M., Graeber B., Schindlmayr G., Managing Energy Risk. Wiley, West Sussex, 2014.

Byers E.A., Hall J.W., Amezaga J.M., Electricity generation and cooling water use: UK pathways to 2050. Global Environmental Change, Vol. 25, 2014.

Carbonara N., Pellegrino R., Public-private partnerships for energy efficiency projects: A win-win model to choose the energy performance contracting structure. Journal of Cleaner Production, Vol. 170, 2018.

Čekanavičius D., Semėnienė F.O., van Ierland E., The Cost of Pollution. [In:] L. Rydén, P. Migula, M. Anderssons (eds.). Environmental Science: Understanding, Protecting and Managing the Environment in the Baltic Sea Region. The Baltic University Press, Uppsala, 2003.

Bibliography

Chan Y., Heer P., Strug K., Onuzo D., Pemble L., Menge J., Technical Assistance Services to Assess the Energy Savings Potentials at National and European Level: Member State Annex Report. European Commission, Directorate-General for Energy, Publications Office, Luxembourg, 2021.

Chien-Chiang L., Chun-Ping C., Energy consumption and GDP revisited: A panel analysis of developed and developing countries. Energy Economics, Vol. 29, no. 6, 2007.

Chiou-Wei S.Z., Ching-Fu C., Zhu Z., Economic growth and energy consumption revisited – Evidence from linear and nonlinear Granger causality. Energy Economics, Vol. 30, no. 6, 2008.

Chodyński A., Odpowiedzialność ekologiczna w proaktywnym rozwoju przedsiębiorstw. Krakowskie Towarzystwo Edukacyjne – Oficyna Wydawnicza AFM, Kraków, 2011.

Chow J., Kopp R.J., Portne P.R., Energy resources and global development. Science, Vol. 302, 2003.

Clean Energy for all Europeans. European Union, 2019.

Cleveland C.J., Morris C.G., (eds.), Dictionary of Energy. Elsevier, Amsterdam, 2005.

Coady D., Parry I., Sears L., Shang B., How large are global energy subsidies? International Monetary Fund, no. 15–105, 2015.

Coase R.H., The problem of social cost. Journal of Law and Economics, Vol. 3, 1960.

Committee on Agriculture and Rural Development, no. 335/10–09–2019, www.sejm.gov.pl, accessed: 30.09.2022.

Communication from the Commission to the European Parliament, The Council, The European Economic and Social Committee and the Committee of the Regions, EU 'Save Energy', European Commission, Brussels, 18.5.2022, COM 240 final, 2022.

Cormio C., Dicorato M., Minoia A., Trovato M., A regional energy planning methodology including renewable energy sources and environmental constraints. Renewable and Sustainable Energy Reviews, Vol. 7, no. 2, 2003.

Cortés-Arcos T., Bernal-Agustín J.L., Dufo-López R. et al., Multi-objective demand response to real-time prices (RTP) using a task scheduling methodology. Energy, Vol. 138, 2017.

de Bruyn S.M., Explaining the environmental Kuznets curve: Structural change and international agreements in reducing sulphur emissions. Environment and Development Economics, Vol. 2, no. 4, 1997.

de Vries B.J.M., van Vuuren D.P., Hoogwijk M.M., Renewable energy sources: Their global potential for the first-half of the 21st century at a global level: An integrated approach. Energy Policy, Vol. 35, 2007.

Debnath K.B., Mourshed M., Forecasting methods in energy planning models. Renewable and Sustainable Energy Reviews, Vol. 88, 2018.

Dessus B., Devin B., Pharabob F., World potential of renewable energies. CNRS—PIRSEM, Paris, 1992.

Devine-Wright P., From backyards to places: Public engagement and the emplacement of renewable energy technologies. [In:] P. Devine-Wright (ed.). Renewable Energy and the Public. From NIMBY to Participation. Routledge, London, 2010.

Diouf B., Pode R., Potential of lithium-ion batteries in renewable energy. Renewable Energy, Vol. 76, 2015.

Dipippo R., Geothermal Power Plants: Principles, Applications, Case Studies and Environmental Iwmpact. Elsevier, Oxford, 2012.

Directive 2001/77/EC of the European Parliament and of the Council of 27 September 2001 on the promotion of electricity produced from renewable energy sources in the internal electricity market. Official Journal L 283, 2001.

Directive 2003/87/EC of the European Parliament and of the Council of 13 October 2003 establishing a scheme for greenhouse gas emission allowance trading within the Community and amending Council Directive 96/61/EC, www.ochronaklimatu.com/dyrektywa-200387we, accessed: 07.09.2022.

Bibliography

Dolhasz M., Fudaliński J., Kosala M., Smutek H., Podstawy zarządzania. Wydawnictwo Naukowe PWN, Warszawa, 2009.

Domańska W., Bochenek D., Dawgiałło U., Gorzkowska E., Hejne J., Kiełczykowska A., Kruszewska D., Labutina Y., Nowakowska B., Sulik J., Wichniewicz A., Wrzosek A., Environment 2020, GUS, Warsaw, 2021.

Dowell G.W.S., Muthulingam S., Will firms go green if it pays? The impact of disruption, cost, and external factors on the adoption of environmental initiatives, Strategic Management Journal, Vol. 38, no. 6, 2017.

Dresner S., Dunne L., Clinch P., Beuermann C., Social and political responses to ecological tax reform in Europe: An introduction to the special issue. Energy Policy, Vol. 34, no. 8, pp. 895–904, 2006.

Drucker P.F., Innovation and Entrepreneurship. Practice and Principles. Harper, New York, 1993.

Drucker P.F., Menedżer skuteczny. Wydawnictwo Akademii Ekonomicznej w Krakowie, Kraków, 1995.

Dyduch J., Rozwój rynku unijnych uprawnień do emisji gazów cieplarnianych. Studia Ekonomiczne, no. 198, 2014.

Dyląg A., Kassenberg A., Szymański W., Civic energy in Poland—Analysis of the state and prospects for development. Institute for Sustainable Development, Warsaw, 2019.

Dyrektywa 2006/32/WE Parlamentu Europejskiego i Rady z dnia 5 kwietnia 2006 roku w sprawie efektywności końcowego wykorzystania energii i usług energetycznych oraz uchylająca dyrektywę Rady 93/76/EWG, 2006.

Dyrektywa 2018/2001 w sprawie promowania stosowania energii ze źródeł odnawialnych – dalej: RED II.

Dyrektywa Parlamentu Europejskiego i Rady (UE) 2018/2001 z dnia 11 grudnia 2018 roku w sprawie promowania stosowania energii ze źródeł odnawialnych.

Ecke J., Steinert T., Bukowski M., Śniegocki A., Polski sektor energetyczny 2050.4 scenariusze, http://forum-energii.eu/files/file_add/file_add-78.pdf, p. 29, accessed: 31.12.2017.

Eduarda F.M., Anabela B., Ligia P. 2010. Grandfathering vs. auctioning in the EU ETS: An experimental study, World congress of environmental and resource economists, Montreal 2010 – Proceedings of the fourth world congress of environmental and resource economists (WCERE 2010), www.webmeets.com/files/papers/WCERE/2010/1503/WCERE%202010%5B1%5D.pdf, accessed: 02.11.2017.

Electric Vehicles in Europe. EEA, Copenhagen, 2016.

Energy import dependency in the European Union (EU-27) in 2020, by country, www.statista.com/, accessed: 20.09.2022.

Energy Management and Gas Supply System in Poland in 2021, GUS, Warsaw, 2022.

Energy to 2050 Scenarios for a Sustainable Future, International Energy Agency. OECD/IEA, Paris, 2003.

Environmental tax statistics, http://ec.europa.eu/eurostat/statistics-explained/index.php/Environmental_tax_statistics, accessed: 07.09.2022.

EU Emissions trading scheme (EU ETS), https://ec.europa.eu/clima/policies/ets_pl, accessed: 07.09.2022.

Fawcett T., Killip G., Re-thinking energy efficiency in European policy: Practitioners' use of 'multiple benefits' arguments, Journal of Cleaner Production, Vol. 210, 2019.

Fischer G., Schrattenholzer L., Global bioenergy potentials through 2050. Biomass and Bioenergy, Vol. 20, no. 3, 2001.

Foster R., Ghassemi M., Cota A., Solar Energy: Renewable Energy and the Environment. CRC Press, Boca Raton, 2010.

Fräss-Ehrfelds C., Renewable Energy Sources: A Chance to Combat Climate Change. Wolters Kluwer, New York, 2009.

Bibliography

Friedmann J., Ogólna teoria rozwoju spolaryzowanego. Przegląd Zagranicznej Literatury Geograficznej, no. 1–2, 1974.

Friedrich R., Voss A., External costs of electricity generation. Energy Policy, Vol. 21, no. 2, 1993.

Fücks R., Zielona rewolucja. Książka i Prasa, Warszawa, 2016.

Fujita R., General remarks Regional revitalization in Fukushima Prefecture. Denki Hyoron, Vol. 100, no. 11, 2015.

Garcia-Casals X., Ferroukhi R., Parajuli B., Measuring the socio-economic footprint of the energy transition. Energy Transitions, Vol. 3, 2019.

Geller H., Harrington P., Rosenfeld A.H., Tanishima S., Unander F., Polices for increasing energy efficiency: Thirty years of experience in OECD countries, Energy Policy, Vol. 34, 2006.

Gillingham K., Rapson D., Wagner G., The rebound effect and energy efficiency policy. Discussion Paper, RFF DP 14–39, Washington, DC, 2014.

Gillingham K., Sweeney J., Market failure and the structure of externalities. [In:] B. Moselle, J. Padilla, R. Schmalensee (eds.). Harnessing Renewable Energy in Electric Power Systems: Theory, Practice, Policy. Earthscan, Earthscan, 2010.

Global EV Outlook 2022, Securing Supplies for an Electric Future. IEA Publications, Paris, 2022.

Global Public Opinion on Nuclear Issues and the IAEA Final Report from 18 Countries, The International Atomic Energy Agency, Vienna, 2005.

Global Trends in Renewable Energy Investment 2017, Frankfurt am Main, http://fsunepcentre. org/sites/default/files/publications/globaltrendsinrenewableenergyinvestment2017.pdf, accessed: 20.11.2017.

Głodziński E., Efektywność ekonomiczna – dylematy definiowania i pomiaru. Zeszyty Naukowe Politechniki Śląskiej, Organizacja i Zarządzanie, no. 73, 2014.

Głodziński E., Efektywność w zarządzaniu projektami. PWE, Warszawa, 2017.

Glossary of Environment Statistics, Studies in Methods, Series F, No. 67. United Nations, New York, 1997. https://stats.oecd.org/glossary/detail.asp?ID=458, accessed: 07.09.2022.

Goldemberg J., Renewable energy. energy efficiency, and emissions trading. [In:] B. Moselle, J. Padilla, R. Schmalensee (eds.). Harnessing Renewable Energy in Electric Power Systems: Theory, Practice, Policy. RFF Press, Washington, DC, 2010.

Goldemberg J., The case for renewable energies. [In:] D. Assmann, U. Laumanns, D. Uh, Earthscan (eds.). Renewable Energy. A Global Revies of Technologies, Policies and Markets. Earthscan, London, 2006.

Gołębiowski S., Krzemień Z., Przewodnik inwestora małej elektrowni wodnej. Fundacja Poszanowania Energii, Warszawa, 1998.

Golecki M.J., Między pewnością a efektywnością. Marginalizm instytucjonalny wobec prawotwórczego stosowania prawa. Wolters Kluwer Polska, Warszawa, 2011.

Gostomczyk W., Zróżnicowanie nakładów pracy i kosztów w sektorze odnawialnych źródeł energii. Roczniki Naukowe Stowarzyszenia Ekonomistów Rolnictwa i Agrobiznesu, Vol. 15, no. 4, 2013.

Graczyk A., Mechanizmy rynkowe w ochronie środowiska jako czynnik zrównoważonego rozwoju. Problemy Ekorozwoju, Vol. 4, no. 1, 2009.

Greenwald B.C., Stiglitz J.E., Externalities in Economies with imperfect information and incomplete markets. The Quarterly Journal of Economics, Vol. 101, no. 2, 1986.

Griffin R.W., Management, Boston Houghton Miffin Company, Boston, 1996.

Griffin R.W., Podstawy zarządzania organizacjami. Wydawnictwo Naukowe PWN, Warszawa, 2021.

Grubb M., Edmonds J., Brink P., Morrison M., The costs of limiting fossil-fuel CO_2 emissions: A survey and analysis. Annual Review of Energy and the Environment, Vol. 18, no. 1, 1993.

Grycan M., Wróblewski Z., Ocena modelowania dynamicznego jako narzędzia do prognozowania zużycia energii elektrycznej przez odbiorców indywidualnych. Prace Instytutu Elektrotechniki, Vol. 63, no. 272, 2016.

Bibliography

Grycan W., Wnukowska B., Wróblewski Z., Modelowanie uwarunkowań zużycia energii elektrycznej regionu. Przegląd Elektrotechniczny, no 2, 2014.

Gulczyński D., Wybrane priorytety i środki zwiększenia efektywności energetycznej. Polityka Energetyczna, Vol. 12, no. 2–2, 2009.

Gungor V.C., Sahin D., Kocak T., Ergut S., Buccella C., Cecati C., Hancke G.P., A survey on smart grid potential applications and communication requirements. IEEE Transactions on Industrial Informatics, Vol. 9, no. 1, 2013.

Gylfason T., Natural resources, education, and economic development. European Economic Review, Vol. 45, no. 4–6, 2001.

Håkansson H., Snehota I., Developing Relationship in Business Networks. Routledge, London, 1995.

Handbook Addressed to Public Sector Entities. KAPE, Warsaw, 2012.

Hansen S.J., Langlois P., Bertoldi P., ESCOs Around the World: Lessons Learned in 49 Countries. The Fairmont Press, Inc., London, 2009.

Hauschild M.Z., Introduction to LCA methodology. [In:] M. Hauschild, R. Rosenbaum, S. Olsen (eds.). Life Cycle Assessment. Springer, Cham, 2018.

Heilmann J., Houle C., Economics of pumping kite generators. [In:] U. Ahrens, M. Diehl, R. Schmehl (eds.). Airborne Wind Energy. Springer Science & Business Media, New York, 2011.

Hitiris T., European Union Economics. Pearson Education, Harlow, 2003.

Horst D., NIMBY or not? Exploring the relevance of location and the politics of voiced opinions in renewable energy siting controversies. Energy Policy, Vol. 35, no. 5, 2007.

Housing Economy and Municipal Infrastructure in 2020, GUS, Warsaw, 2021.

https://energiapress.pl, accessed: 30.09.2022.

https://zielonerozwiazania.pl, accessed: 1.10.2022.

Hu Y., Guo D., Wang M., Zhang X., Wang S., The relationship between energy consumption and economic growth: Evidence from china's industrial sectors. Energies, Vol. 8, 2015.

Iannuzzi A., Industry Self-Regulation and Voluntary Environmental Compliance. Taylor & Francis Group LLC, Boca Raton, 2002.

Inglesi-Lotz R., The impact of renewable energy consumption to economic growth: A panel data application. Energy Economics, Vol. 53, 2016.

International Energy Agency, World Energy Outlook. OECD/IEA, Paris, 2009.

IRENA, Roadmap for a Renewable Energy Future, 2016 edition. International Renewable Energy Agency (IRENA), Abu Dhabi, 2016.

IRENA, Global Energy Transformation. A Roadmap to 2050. IRENA, Abu Dhabi, 2018.

IRENA, Renewable Energy and Jobs Annual Review. IRENA, Abu Dhabi, 2021.

Jakubczyk Z., Prekursorskie teorie rozwoju krajów zacofanych. [In:] B. Fiedor, K. Kociszewski (eds.). Ekonomia rozwoju. Wydawnictwo Uniwersytetu Ekonomicznego we Wrocławiu, Wrocław, 2010.

Jankowska E., Środowiskowa krzywa Kuznetsa w dekarbonizacji europejskich gospodarek. Studia Ekonomiczne. Zeszyty Naukowe Uniwersytetu Ekonomicznego w Katowicach, no. 289, 2016.

Janowicz L., Biomasa w Polsce. Energetyka i ekologia, no. 8, 2006.

Jarno K., Przyczynek do dyskusji nad zasadnością włączenia do europejskiego systemu handlu emisjami sektorów dotychczas nieobjętych systemem. Ruch Prawniczy, Ekonomiczny i Socjologiczny, no. 3, 2016.

Jarosiński K., Innovations in the public sector and their impact on the social-economic development processes. [In:] K. Jarosiński (ed.). Making the 21st Century Cities. CeDeWu, Warszawa, 2015.

Jarosiński K., Maśloch G., Opałka B., Grzymała Z., Financing and Management of Public Sector Investments on Local and Regional Levels. PWN, Warszawa, 2015.

Jastrzębski J., Mroczek K., Ronald Harry Coase, 1910–2 Gospodarka Narodowa, no 3, 2014.

Bibliography

Jebaraj S., Iniyan S., A review of energy models. Renewable and Sustainable Energy Reviews, Vol. 10, no. 4, 2006.

Johansson T.B., McCormick K., Neij L., Turkenburg W., The potentials of renewable energy. International Conference for Renewable Energies, Bonn, 2004.

Johnstone N., Haščič I., Popp D., Renewable energy policies and technological innovation: Evidence Based on patent counts. Environmental and Resource Economics, Vol. 45, no. 1, 2010.

Jones G., Bouamane L., Power from Sunshine: A Business History of Solar Energy. Harvard Business School, Boston, 2012.

Jordan-Korte K., Government Promotion of Renewable Energy Technologies: Policy Approaches and Market Development in Germany, the United States, and Japan. Springer Science & Business Media, London, 2011.

Kampa M., Castanas E., Human health effects of air pollution. Environmental Pollution, Vol. 151, no. 2, 2008.

Kander A., Malanima P., Warde P., Power to the People. Princeton University Press, New Jersey, 2013.

Karabiber O.A., Xydis G., Forecasting day-ahead natural gas demand in Denmark. Journal of Natural Gas Science and Engineering, Vol. 76, 2020.

Kardooni R., Yusoff S.B., Kari F.B., Moeenizadeh L., Public opinion on renewable energy technologies and climate change in Peninsular Malaysia. Renewable Energy, Vol. 116, 2018.

Karkour S., Ichisugi Y., Abeynayaka A., Itsubo N., External-cost estimation of electricity generation in g20 countries: Case study using a global life-cycle impact-assessment method. Sustainability, Vol. 12, no. 5, 2020.

Kassenberg A., Szymański W., Energetyka obywatelska w Polsce – analiza stanu i perspektywy rozwoju. Wyd. Instytut na rzecz ekorozwoju, Warszawa, 2019.

Kilduff M., Tsai W., Social Networks and Organizations. Sage, London, 2013.

Kim J., Park S.Y., Lee J., Do people really want renewable energy? Who wants renewable energy?: Discrete choice model of reference-dependent preference in South Korea. Energy Policy, Vol. 120, 2018.

Kirchhoff B.A., Organization effectiveness measurement and policy research. Academy of Management Review, Vol. 2, no. 3, 1977.

Klein N., To zmienia wszystko. Kapitalizm kontra klimat. Muza, Warszawa, 2016.

Kost C., Mayer J.N., Thomsen J., Hartmann N., Senkpiel Ch., Philipps S., Nold S., Lude S., Saad N., Schlegl T., Levelized Cost of Electricity Renewable Energy Technologies. Fraunhofer Institute for Solar Energy Systems ISE, Freiburg im Breisgau, 2013.

Kotler P., Marketing. Analiza, planowanie, wdrażanie i kontrola. Wydawnictwo Gebethner & Ska, Warszawa, 1994.

Kotler P., Twarzą w twarz z kapitalizmem. Realne rozwiązania dla niezdrowego systemu ekonomicznego. MTBiznes, Warszawa, 2016.

Krajowy Plan Rozwoju Mikroinstalacji Odnawialnych Źródeł Energii do roku 2Instytut Energetyki Odnawialnej, Warszawa, 2015.

Krawiec F., Rola odnawialnych źródeł energii w rozwiązywaniu globalnego kryzysu energetycznego. [In:] F. Krawiec (ed.). Odnawialne źródła energii w świetle globalnego kryzysu. Difin, Warszawa, 2010.

Krzyszkowska J., Energetyka obywatelska, przewodnik dla samorządów, po inwestycjach w odnawialne źródła energii i efektywności energetycznej, CEE Bankwatch Network., Warszawa, 2015.

Księżopolski K., Maśloch G., Kotlewski D., Nowe zielone otwarcie w energetyce Europy Środkowo-Wschodniej. Raport SGH i Forum Ekonomicznego, 2021.

Księżopolski K., Maśloch G., Kotlewski D., Morawiecka M., Sektor energetyczny w dobie szoków cenowych i wojny hybrydowej. Raport SGH i Forum Ekonomicznego w Warszawie, Warszawa, 2022.

Kudełko M., Pękala E., Ekologiczna reforma podatkowa – Wyzwania i ograniczenia. Problemy Ekologii, Vol. 12, no. 1, 2008.

Kudełko M., Znaczenie analizy systemowej w prognozowaniu rozwoju sektorów paliwowo-energetycznych. Polityka Energetyczna, Vol. 8, 2005.

Kuklo K., Energia słoneczna – dostępność, możliwości wykorzystania, korzyści. [In:] Energia odnawialna. Jak z niej korzystać? Podlaska Fundacja Rozwoju Regionalnego, Białystok, 2007.

Landa T., A Theory of the ethnically homogeneous middleman group: An institutional alternative to contract law. The Journal of Legal Studies, Vol. 10, no. 2, 1981.

Lee M.K., Park H., Noh J., Painuly J.P., Promoting energy efficiency financing and ESCOs in developing countries: Experiences from Korean ESCO business. Journal of Cleaner Production, Vol. 11, no. 6, 2003.

Lee P., Lam P.T.I., Lee W.L., Risks in energy performance contracting (EPC) projects. Energy and Buildings, Vol. 92, 2015.

Li M., Wang W., De G., Ji X., Tan Z., Forecasting carbon emissions related to energy consumption in beijing-tianjin-hebei region based on grey prediction theory and extreme learning machine optimized by support vector machine algorithm. Energies, Vol. 11, 2018.

Li Y., The case analysis of the scandal of enron. International Journal of Business and Management, Vol. 5, no. 10, 2010.

Long-term Renovation Strategy. Supporting the Renovation of the National Construction Resource, Warsaw, 2022. Annex to Resolution No. 23/2022 of the Council of Ministers of February 9, 2022.

Lund H., Choice awareness: The development of technological and institutional choice in the public debate of Danish energy planning. Journal of Environmental Policy and Planning, Vol. 2, 2000.

Lund H., Renewable Energy Systems: A Smart Energy Systems Approach to the Choice and Modeling of 100% Renewable Solutions. Academic Press, Oxford, 2004.

Lund H., Mathiesen B.V., Energy system analysis of 100% renewable energy systems. The case of Denmark in years 2030 and 2Energy, Vol. 34, no. 5, 2009.

Luukkanen J., Vehmas J., Mustonen S., Allievi F., Karjalainen A., Värttö M., Ahoniemi M., Finnish Energy Industries – Energy Scenarios and Visions for the Future. Finland Futures Research Centre, Turku School of Economics, Turku, 2019.

Maciejewski J., Procesy społeczne a zmiany świadomości w kwestiach bezpieczeństwa. Acta Universitatis Wratislaviensis, No 3096, Socjologia XLV, Wrocław, 2009.

Maedas A., The emergence of market power in emission rights markets: The role of initial permit distribution. Journal of Regulatory Economics, Vol. 24, no. 3, 2003.

Malara Z., Przedsiębiorstwo w globalnej gospodarce. Wyzwania współczesności. PWN, Warszawa, 2006.

Maśloch G., Budowa autonomicznych regionów energetycznych w Polsce—utopia czy konieczność? Studia Prawno-Ekonomiczne, Tom CVI, Łódź, 2018.

Maśloch G., Gospodarowanie odpadami komunalnymi w aspekcie wyzwań wynikających z realizacji koncepcji zrównoważonego rozwoju (wybrane problemy). Studia i Prace. Kolegium Zarządzania i Finansów. SGH, Zeszyt naukowy 138, Warszawa, 2014.

Maśloch G., Gospodarowanie odpadami komunalnymi w Unii Europejskiej. [In:] Z. Grzymała, G. Maśloch, M. Goleń, E. Górnicki (eds.). Racjonalizacja gospodarki odpadami komunalnymi w Polsce w świetle zmian Ustawy o utrzymaniu czystości i porządku w gminach. SGH, Warszawa, 2013.

Maśloch G., Ocena i finansowanie komunalnych projektów inwestycyjnych współfinansowanych ze środków UE. [In:] Z. Grzymała (ed.). Podstawy ekonomiki i zarządzania w gospodarce komunalnej. Oficyna Wydawnicza SGH, Warszawa, 2011.

Maśloch G., Potencjał (zasoby) energii odnawialnej w Polsce oraz ich regionalne zróżnicowanie. [In:] G. Maśloch (ed.). Zarządzanie energetyką lokalną w aspekcie wykorzystania odnawialnych źródeł energii. SGH, Warszawa, 2012.

Bibliography

Maśloch G., Rola jednostek samorządu lokalnego w zakresie ograniczania niskiej emisji (w aspekcie perspektywy finansowej Unii Europejskiej na lata 2014–2020). Studia i Prace Kolegium Zarządzania i Finansów. Szkoła Główna Handlowa, Vol. 146, 2015.

Maśloch G., Rola i znaczenie prosumenta we wdrażaniu zasady zrównoważonego rozwoju (zmiana paradygmatu rozwoju energetyki). [In:] R. Bartkowiak, P. Wachowiak (eds.). Nowe paradygmaty w naukach ekonomicznych. Oficyna Wydawnicza SGH, Warszawa, 2016.

Maśloch G., Udział energii ze źródeł odnawialnych w bilansie energetycznym państw Unii Europejskiej. [In:] Zarządzanie energetyką lokalną w aspekcie wykorzystania odnawialnych źródeł energii. SGH, Warszawa, 2012.

Maśloch P., Maśloch G., Kuźmiński Ł., Wojtaszek H., Miciuła I., Autonomous energy regions as a proposed choice of selecting selected eu regions – Aspects of their creation and management. Energies, Vol. 13, no. 23, 2020.

Masur J.S., Posner E.A., Toward a Pigovian State. Coase-Sandor Working Paper Series in Law and Economics, No. 716. The University of Chicago, Chicago, 2015.

Matusiak B.E., Pamuła A., Zieliński J.S., Narzędzia ICT w sterowaniu zachowaniem klienta w inteligentnych sieciach energetycznych. [In:] R. Knosala (ed.). Komputerowo zintegrowane zarządzanie. Oficyna Wydawnicza Polskiego Towarzystwa Zarządzania Produkcją, Opole, 2011.

Matuszczyk P., Popławski T., Flasza J., Potencjał i możliwości energii promieniowania elektromagnetycznego Słońca. Przegląd Elektrotechniczny, no. 1, 2015.

McMichael A.J., Global warming, ecological disruption and human health: The penny drops. The Medical Journal of Australia, Vol. 154, 1991.

Melikoglu M., Current status and future of ocean energy sources: A global review. Ocean Engineering, Vol. 148, 2018.

Mendonça M., Lacey S., Hvelplund F., Stability, participation and transparency in renewable energy policy: Lessons from Denmark and the United States. Policy and Society, Vol. 27, no. 4, 2009.

Menten F., Chèze B., Patouillard L., Bouvart F., A review of LCA greenhouse gas emissions results for advanced biofuels: The use of meta-regression analysis. Renewable and Sustainable Energy Reviews, Vol. 26, 2013.

Mianowski A., Kryzys przemysłowego rozwoju Czystych Technologii Węglowych w Polsce, 2012, http://chemia.wnp.pl/kryzys-przemyslowego-rozwoju-czystych-technologii-weglowych-w-polsce,-7539_2_0_0.html, accessed: 13.09.2022.

Ministry of Climate and Environment, Energy Policy of Poland Until 2040 (EPP2040). Ministry of Climate and Environment, Warsaw, 2021.

Mikuła B., Organizacje oparte na wiedzy. Wydawnictwo Akademii Ekonomicznejw Krakowie, Kraków, 2006.

Mohtasham J., Renewable energies. Energy Procedia, Vol. 74, 2015.

Morgül Tumbaz M.N., İpek M., Energy demand forecasting: Avoiding multi-collinearity. Arabian Journal for Science and Engineering, Vol. 46, 2021.

Moriarty P., Honnery D., What is the global potential for renewable energy? Renewable and Sustainable Energy Reviews, Vol. 16, 2012.

Motowidlak T., Programy DSR instrumentem poprawy bezpieczeństwa dostaw energii elektrycznej. Przegląd Naukowo-Metodyczny, no. 1(34), 2017.

Motowidlak U., Polityka Unii Europejskiej na rzecz zwiększenia efektywności ekonomicznej i środowiskowej transportu. Cz. 1: Poprawa efektywności energetycznej transportu. Logistyka, no. 3, 2014.

Nel P.J.C., Booysen M.J., van der Merwe B., Energy perceptions in South Africa: An analysis of behaviour and understanding of electric water heaters. Energy for Sustainable Development, Vol. 32, 2016.

Nesta L., Vona F., Nicolli F., Environmental policies, competition and innovation in renewable energy. Journal of Environmental Economics and Management, Vol. 67, no. 3, 2014.

Bibliography

Niemyski M., Tatarewicz I., Ocena zasobów odnawialnych źródeł energii możliwych technicznie i ekonomicznie do wykorzystania w celu produkcji energii elektrycznej. Raport cząstkowy 4, Warszawa, 2008.

Nilsson B.E., Using Voluntary Agreements in Environmental Policy: A Reinforcement of the Dialogue with Industry. The Nordic Council of Ministers, Stockholm, 1998.

Owen A.D., Environmental externalities, market distortions and the economics of renewable energy technologies. The Energy Journal, Vol. 25, no. 3, 2004.

Pablo-Romero M. del P., Sánchez-Braza A., Productive energy use and economic growth: Energy, physical and human capital relationships. Energy Economics, Vol. 49, 2015.

Paish O., Small hydro power: Technology and current status. Renewable and Sustainable Energy Reviews, no. 6, 2002.

Pamuła A., Papińska-Kacperek J., Inteligentne domy i inteligentne sieci energetyczne jako elementy infrastruktury smart city. Studia Informatica, no. 29, 2012.

Panwar N.L., Kaushik S.C., Kotharia S., Role of renewable energy sources in environmental protection: A review. Renewable and Sustainable Energy Reviews, Vol. 15, no. 3, 2011.

Pasour E.C., A Further note on the measurement of efficiency and economies of farm size. Journal Agriculture Economic, no. 32, 1981.

Pasqualetti M.J., Social barriers to renewable energy landscapes. Geographical Review, Vol. 101, no. 2, 2011.

Patronen J., Kaura E., Torvestad C., Nordic Heating and Cooling: Nordic Approach to EU's Heating and Cooling Strategy. Nordic Council of Ministers, Rosendahl, 2017.

Pawel I., The cost of storage – How to calculate the levelized cost of stored energy (LCOE) and applications to energy generation. Energy Procedia, Vol. 46, 2014.

Pearce D., Turner K., Economics of Natural Resources and the Environment. Harvester Whateshaw, London 1990.

Pfenninger S., Hawkes A., Keirstead J., Energy systems modeling for twenty-first century energy challenges. Renewable and Sustainable Energy Reviews, Vol. 33, 2014.

Pieńkowski D., Paradoks Jevons'a a konsumpcja energii w Unii Europejskiej. Problemy Ekorozwoju, Vol. 7, no. 1, 2012.

Pietraszewski M., Funkcjonowanie przedsiębiorstwa w warunkach gospodarki rynkowej. eMPi2, Poznań, 2004.

Polimeni J.M., Mayumi K., Giampietro M., Alcott B., The Myth of Resource Efficiency: The Jevons Paradox. Earthscan, New York, 2009.

Polityka energetyczna Polski do 2030 roku. Załącznik do uchwały nr 202/2009 Rady Ministrów z dnia 10 listopada 2009 r.

Pomaskow J., Twierdzenie Coase'a a narodziny ekonomicznej analizy prawa. Studia Ekonomiczne. Zeszyty Naukowe Uniwersytetu Ekonomicznego w Katowicach, no. 259, 2016.

Popp J., Lakner Z., Harangi-Rákos M., Fári M., The effect of bioenergy expansion: Food, energy, and environment. Renewable and Sustainable Energy Reviews, no. 32, 2014.

Pszczołowski T., Mała encyklopedia prakseologii i teorii organizacji. Zakład Narodowy im. Ossolińskich, Wrocław–Warszawa–Kraków, 1978.

Pyszka A., Istota efektywności. Definicje i wymiary. Studia Ekonomiczne. Zeszyty Naukowe Uniwersytetu Ekonomicznego w Katowicach, no. 230, 2015.

Pytliński Ł., Węgiel, stare piece i brak ocieplenia. Systemy grzewcze i izolacja termiczna w sektorze domów jednorodzinnych w Polsce. Raport z badań. [In:] Efektywność energetyczna w Polsce. Przegląd, Instytut Ekonomii Środowiska, Kraków, 2013.

Rahman A., Srikumar V., Simith A., Predicting electricity consumption for commercial and residential buildings using deep recurrent neural networks. Applied Energy, Vol. 212, 2018.

Rana M.S., AlHumaidan F.S., Statistical data on worldwide coal reserves, production, consumption, and future demand. [In:] M.R. Riazi, R. Gupta (eds.). Coal Production and Processing Technology. Taylor & Francis Group, Boca Raton, 2016.

Bibliography

Ranosz R., Organizacja i handel uprawnieniami do emisji CO2. Polityka Energetyczna, Vol. 11, no. 2, 2008.

Rasolomampionona D., Robak S., Chmurski P., Tomasik G., Przegląd istniejących mechanizmów DSR stosowanych na rynkach energii elektrycznej. Rynek Energii, no. 4, 2010.

Rathnayaka A.J.D., Potdar V.M., Hussain O., Dillon T., Identifying prosumer's energy sparing behaviours for forming optimal prosumer communities. International Conference on Cloud and Service Computing, Hong Kong, 2011.

Rehman S., Cai Y., Fazal R., Das Walasai G., Mirjat N., An integrated modeling approach for forecasting long-term energy demand in Pakistan. Energies, Vol. 10, 2017.

Renewable Energy in Europe: Building Markets and Capacity. European Renewable Energy Council, Brussels, 2004.

Renewable Energy Sources and Climate Change Mitigation, Summary for Policymakers and Technical Summary. Special Report of the Intergovernmental Panel on Climate Change. Intergovernmental Panel on Climate Change, Potsdam, 2012.

Renewables 2022 Global Status Report. REN21, Paris, 2022.

Renewables for heating and cooling and EU security of supply: Save over 20 billion euro annually in reduced fossil fuel imports, p. 5, 2014, www.egec.org.

Resch G., Held A., Faber T. et al., Potentials and prospects for renewable energies at global scale. Energy Policy, Vol. 11, no. 36, 2008.

Review of Evidence on Health Aspects of Air Pollution – REVIHAAP Project. World Health Organization Europe, Copenhagen, 2013.

Rhodes J.D., King C., Gulen G. et al., A geographically resolved method to estimate levelized power plant costs with environmental externalities. Energy Policy, Vol. 102, 2017.

Richter M., Business model innovation for sustainable energy: German utilities and renewable energy. Energy Policy, Vol. 62, 2013.

Rifkin J., The Third Industrial Revolution: How Lateral Power is Transforming Energy, The Economy, and the World. Palgrave Macmillan, New York, 2011.

Robert K.W., Parris T.M., Leiserowitz A.A., What is sustainable development? Goals, indicators, values, and practice. Journal Environment: Science and Policy for Sustainable Development, Vol. 47, no. 3, 2005.

Rosiek K., Ekonomia społeczna a ochrona środowiska. [In:] M. Frączka, J. Hausner, S. Mazur (eds.). Wokół ekonomii społecznej, Małopolska Szkoła Administracji Publicznej. Uniwersytet Ekonomiczny w Krakowie, Kraków, 2012.

Rosser A., The Political Economy of the Resource Curse: A Literature Survey. Institute of Development Studies, Brighton, 2006.

Rumianowska I., Normatywne a ekonomiczne ujęcie regulacji ekologicznych, ze szczególnym uwzględnieniem roli grup interesu w ich kształtowaniu. Ekonomia i Prawo. Zawodności rynku, zawodności państwa, etyka zawodowa, 2010.

Ruszkowski P., Problemy polskiej energetyki w perspektywie socjologicznej. Energetyka— Społeczeństwo – Polityka, no. 1, 2015.

Sachs J.D., Warner A.M., The big push, natural resource booms and growth. Journal of Development Economics, Vol. 59, 1999.

Samuelson W.F., Nordhaus W.D., Ekonomia. Dom Wydawniczy REBIS, Poznań, 1991.

Schlomann B., Rohde C., Plötz P., Dimensions of energy efficiency in a political context. Energy Efficiency, Vol. 8, 2015.

Schumpeter J.A., Teoria rozwoju gospodarczego. PWN, Warszawa, 1960.

Sens L., Neuling U., Kaltschmitt M., Capital expenditure and levelized cost of electricity of photovoltaic plants and wind turbines – Development by 2050. Renewable Energy, Vol. 185, 2022.

Seyfang G., The New Economics of Sustainable Consumption. Seeds of Change. Palgrave Macmillan, New York, 2009.

Bibliography

Shang T., Zhang K., Liu P., Chen Z., Li X., Wu X., What to allocate and how to allocate? Benefit allocation in shared savings energy performance contracting projects. Energy, Vol. 91, 2015.

Shove E., What is wrong with energy efficiency? Building Research & Information, Vol. 46, no. 7, 2018.

Sidorczuk-Pietraszko E., Wpływ instalacji odnawialnych źródeł energii na tworzenie miejsc pracy w wymiarze lokalnym. Ekonomia i Środowisko, Vol. 3, no. 54, 2015.

Sims S., Dent P., Oskrochi G.R., Modelling the impact of wind farms on house prices in the UK. International Journal of Strategic Property Management, no. 12, 2008.

Skoczkowski T., Bielecki S., Konieczność zapewnienia interesów odbiorców końcowych w procesie budowy sieci inteligentnych. Przegląd Elektrotechniczny, no. 1, 2015.

Skoczkowski T., Bielecki S., Środki poprawy efektywności energetycznej w przemyśle i ich ocena. Energetyka, no. 1, 2016.

Skrzypek E., Efektywność ekonomiczna jako ważny czynnik sukcesu organizacji. Prace Naukowe Uniwersytetu Ekonomicznego we Wrocławiu, no. 262, 2012.

Śleszyński J., Koncepcja ekologicznej reformy podatkowej. [In:] W. Stodulski (ed.). Ekologiczna reforma podatkowa. System podatkowy jako instrument zrównoważonego rozwoju w Polsce w pierwszej dekadzie XXI wieku. Instytut na rzecz Ekorozwoju, Warszawa, 2001.

Śleszyński J., Podatki środowiskowe i podział na grupy podatków według metodyki eurostatu, optimum. Studia Ekonomiczne, Vol. 3, no. 69, 2014.

Słonimiec J., Szatkowska P., Stępień N., Urban J., Dobosz S., Biernacki G., Ji H.Y., Fulfilment of goals of the integrated product policy in Poland in comparison to other European union member states. Management Systems in Production Engineering, Vol. 4, no. 20, 2015.

Socolow R., Environment-respectful global development of the energy system. Perspectives in Energy, Vol. 1, no. 1, 1991.

Solarin S.A., Towards sustainable development in developing countries: Aggregate and disaggregate analysis of energy intensity and the role of fossil fuel subsidies. Sustainable Production and Consumption, Vol. 24, 2020.

Song Ch., Global challenges and strategies for control, conversion and utilization of CO2 for sustainable development involving energy, catalysis, adsorption and chemical processing. Catalysis Today, no. 115, 2006.

Sovacool B.K., Dworkin M.H., Global Energy Justice, Problems, Principles and Practices. Cambridge University Press, Cambridge, 2014.

Staliński A., Perspektywy dla inwestowania w biomasę wykorzystywaną do produkcji energii elektrycznej w Polsce—Problem dostępności ziemi na przykładzie uprawy wierzby wiciowej. Studia Oeconomica Posnaniensia, Vol. 4, no. 6, 2016.

Staniszewska L., Materialne i proceduralne zasady stosowane przy wymierzaniu administracyjnych kar pieniężnych. [In:] M. Błachucki (ed.). Administracyjne kary pieniężne w demokratycznym państwie prawa. Biuro Rzecznika Praw Obywatelskich, Warszawa, 2015.

Stern D.I., Common M.S., Barbier E.B., Economic growth and environmental degradation: The environmental Kuznets curve and sustainable development. World Development, Vol. 24, no. 7, 1996.

Stern D.I., The rise and fall of the environmental Kuznets curve. World Development, Vol. 32, no. 8, 2004.

Stober I., Bucher K., Geothermal Energy: From Theoretical Models to Exploration and Development. Springer, Berlin, 2013.

Stoner J.A.F., Freeman R.E., Gilbert D.R., Kierowanie. PWE, Warszawa, 2001.

Storm S., Financialization and economic development: A debate on the social efficiency of modern finance. Development and Change, Vol. 49, no. 2, 2018.

Suganthi L., Samuel A.A., Energy models for demand forecasting – A review. Renewable and Sustainable Energy Reviews, Vol. 16, 2012.

Bibliography

Suri V., Chapman D., Economic growth, trade and energy: Implications for the environmental Kuznets curve. Ecological Economics, Vol. 25, no. 2, 1998.

Swilling M., Contested futures: Conceptions of the next long-term development cycle. [In:] M. McIntosh (ed.). The Necessary Transition: The Journey towards the Sustainable Enterprise Economy. Routledge, New York, 2013.

Szczerbowski R., Modelowanie systemów energetycznych. Electrical Engineering, no. 78, 2014.

Szczukowski S., Tworkowski J., Zmiany w produkcji i wykorzystaniu biomasy w Polsce. [In:] Praktyczne aspekty wykorzystania odnawialnych źródeł energii. Plan energetyczny województwa podlaskiego. Podlaska Fundacja Rozwoju Regionalnego, Białystok, 2006.

Szulecki K., Szwed D., Społeczne aspekty OŹE: którędy do energetycznej demokracji? [In:] K.M. Księżopolski, K.M. Pronińska, A.E. Sulowska (eds.). Odnawialne źródła energii w Polsce Wybrane problemy bezpieczeństwa, polityki i administracji. Dom Wydawniczy ELIPSA, Warszawa, 2013.

Tan K.M., Thanikanti S., Babu T.S., Ramachandaramurthy V.K., Kasinathan P., Solanki S.G., Raveendran S.K., Empowering smart grid: A comprehensive review of energy storage technology and application with renewable energy integration. Journal of Energy Storage, Vol. 39, 2021.

Telega I., Szanse i zagrożenia stosowania podatków ekologicznych w praktyce gospodarczej. [In:] A. Nalepka, A. Ujwara-Gil (eds.). Organizacje komercyjne i niekomercyjne wobec wzmożonej konkurencji oraz wzrastających wymagań konsumentów. WSB-NLU, Nowy Sącz, 2010.

The Unpaid Health Bill. HEAL, Brussel, 2013.

Tietenberg T.H., Emissions Trading, an Exercise in Reforming Pollution Policy. Resources for the Future, Washington, DC, 1985.

Tiwari G.N., Mishra R.K., Advanced Renewable Energy Sources. RSC Publishing, Cambridge, 2012.

Tobin J., On the efficiency of the financial system. Lloyds Bank Review, no. 153, 1984.

Torvik R., Natural resources, rent seeking and welfare. Journal of Development Economics, Vol. 67, no. 2, 2002.

Turkenburg W.C., Renewable energy technologies. [In:] J. Goldemberg (ed.). World Energy Assessment. UNDP, Washington, DC, 2000.

Twidell J., Weir T., Renewable Energy Resources. Routledge, New York, 2015.

Twidell J., Weir T., Renewable Energy Resources. Taylor & Francis Group, New York, 2003.

Tymiński J., Wykorzystanie odnawialnych źródeł energii w Polsce do 2030 roku. IBMiR, Warszawa, 1997.

U.S. Energy Information Administration (EIA), World Energy Projection System (2021), run r_210719.163829; and EIA, Annual Energy Outlook 2021, February, www.eia.gov/aeo.

Ujwary-Gil A., Koncepcja zasobowej teorii przedsiębiorstwa – Całościowe ujęcie i kierunek dalszych badań. Przegląd Organizacji, no. 6, 2009.

Van Egteron H., Weber M., Marketable permits, market power, and cheating. Journal of Environmental Economics and Management, Vol. 30, no. 2, 1996.

Von Weizsacker E.U., Factor five: A global imperative. [In:] E.U. von Weizsacker, Ch. Hargroves, M.H. Smith, C. Desha, P. Stasinopoulos (eds.). Factor Five: Transforming the Global Economy through 80% Improvements in Resource Productivity. Earthscan, Sterling, 2009.

Walkowska K., Energy Efficiency in Poland in Years 2010–2020. GUS, Warsaw, 2022.

Walkowska K., Berent-Kowalska G., Peryt S., Dziedzina K., Jurgaś A., Kacprowska J., Gilecki R., Parciński G., Boczek-Gizińska R., Szymańska M., Zatorska M., Żarek E., Pawelczyk M., Moskal I, Energy Statistics in 2019 and 2020. GUS, Warsaw, 2021.

Wiatkowski M., Rosik-Dulewska Cz., Stan obecny i możliwości rozwoju energetyki wodnej w województwie opolskim. Woda-Środowisko-Obszary Wiejskie, Vol. 12, pp. 2–38, 2012.

Wilts H., O'Brien M., A policy mix for resource efficiency in the EU: Key instruments, challenges and research needs. Ecological Economics, Vol. 155, 2019.

Wójcicki Z., Poszanowanie energii i środowiska w rolnictwie i na obszarach wiejskich. Infrastruktura i Ekologia Terenów Wiejskich, no. 2–1, 2006.

Woodcock J., Banister D., Edwards P., Prentice A.M, Roberts I., Energy and transport, The Lancet, Vol. 370, no. 9592, 2007.

World Energy Council, Deciding the Future: Energy Policy Scenarios to 2050. World Energy Council, London, 2007.

World energy resources solar 2016, www.worldenergy.org/wp-content/uploads/2017/03/WEResources_Solar_2016.pdf. pp. 7–8, accessed: 02.01.2018.

Wüstenhagen R., Wolsink M., Bürer J., Social acceptance of renewable energy innovation: An introduction to the concept. Energy Policy, Vol. 35, no. 5, 2007.

www.eia.gov/energyexplained/index.cfm?page=renewable_home, accessed: 01.03.2017.

www.ey.com, accessed: 28.09.2022.

www.iea.org, accessed: 28.09.2022.

www.nationalgeographic.org/encyclopedia/non-renewable-energy/, accessed: 04.04.2017.

Xiaoling O., Boqiang L., Levelized cost of electricity (LCOE) of renewable energies and required subsidies in China. Energy Policy, Vol. 70, no. C, 2014.

Yamamoto H., Yamaij K., Fujino J., Evaluation of bioenergy resources with a global land use and energy model formulated with SD technique. Applied Energy, Vol. 63, 1999.

Yu B., Xu L.Y., Study of eco-compensation in hydropower development in China. Procedia Environmental Sciences, Vol. 13, 2012.

Zalega T., Ekonomia ewolucyjna jako jeden z nurtów współczesnej ekonomii – zarys problematyki. Studia i Materiały, no. 19, 2015.

Zdyb M., Istota decyzji, Wyd. UMCS, Lublin, 1993.

Zieliński M., Efektywność – ujęcie ekonomiczne i społeczne. Zeszyty Naukowe Politechniki Śląskiej, Organizacja i Zarządzanie, no. 66, 2013.

Życzyńska A., Wykorzystanie audytu oraz świadectwa energetycznego budynku przy zarządzaniu nieruchomością. Budownictwo i Architektura, Vol. 12, no. 4, 2013.

Żylicz T., The economics international environmental cooperation, polish studies in economics. PL Academic Research, Vol. 3, 2015.

Index

B

biomass, 3, 46, 85, 86, 88, 91, 92, 93, 95, 97, 107, 112, 113, 116, 117, 118, 126, 132, 146

C

capital, 23, 32, 38, 43, 44, 45, 57, 58, 62, 74, 87, 91, 99, 103, 104, 110, 118, 134, 135, 137, 138, 139, 140, 141, 142, 144

change management, 81, 83

civic energy, 1, 2, 3, 4, 16, 22, 24, 25, 46, 47, 72, 73, 75, 77, 81, 91, 93, 95, 99, 101, 109, 119, 120, 121, 123, 124, 125, 127, 128, 129, 130, 131, 133, 134, 135, 136, 137, 138, 139, 140, 141, 142, 143, 144, 145, 146, 147

conventional sources, 7, 44, 57

costs in the energy sector, 3, 45, 59

D

development, 1, 2, 3, 4, 5, 7, 9, 10, 11, 12, 13, 14, 15, 16, 17, 23, 24, 25, 27, 29, 32, 33, 34, 35, 36, 38, 39, 40, 44, 46, 47, 48, 51, 52, 58, 59, 60, 61, 62, 63, 67, 69, 70, 72, 73, 74, 75, 76, 77, 78, 79, 80, 82, 88, 89, 90, 91, 92, 93, 94, 95, 98, 99, 101, 102, 103, 104, 109, 112, 116, 118, 119, 120, 121, 123, 124, 125, 128, 129, 130, 131, 132, 133, 134, 135, 136, 137, 138, 139, 140, 141, 142, 143, 145, 146, 147

E

economy, 1, 3, 7, 9, 11, 12, 13, 16, 20, 22, 23, 28, 29, 32, 33, 35, 38, 42, 43, 44, 46, 47, 52, 55, 63, 64, 70, 71, 72, 73, 76, 77, 78, 89, 91, 94, 96, 101, 103, 105, 106, 109, 110, 118, 119, 120, 137, 141, 146, 147

electricity, 1, 15, 17, 18, 19, 20, 21, 22, 23, 37, 40, 41, 52, 61, 71, 73, 77, 78, 79, 80, 86, 87, 88, 90, 98, 103, 106, 107, 108, 109, 110, 112, 113, 114, 119, 120, 121, 143, 145

energy consumption, 7, 8, 9, 10, 13, 14, 17, 18, 19, 20, 32, 34, 36, 37, 39, 41, 62, 63, 64, 72, 97, 100, 105, 106, 109, 110, 111, 112, 118, 119

energy cooperative, 73, 121, 125, 128, 130, 131, 134, 135, 137, 143, 145, 146

energy sector, 1, 2, 3, 5, 10, 11, 12, 15, 16, 23, 25, 30, 33, 35, 43, 45, 48, 54, 59, 64, 73, 76, 81, 89, 91, 92, 93, 94, 96, 100, 102, 103, 105, 109, 115, 117, 118, 119, 120, 127, 132, 133, 134, 135, 137, 139, 145, 146, 147

energy storage, 15, 16, 17, 21, 22, 73, 142

enterprises, 2, 13, 23, 25, 80, 110, 134, 135

external costs, 13, 41, 44, 50, 51, 52, 55, 59, 63, 65, 67

European Union, 6, 7, 58, 60, 88, 109, 117, 120

F

forecasting, 33, 34, 35, 46

G

geothermal energy, 85, 86, 87, 92, 97, 113, 116, 117

global, 4, 5, 6, 7, 8, 10, 13, 22, 24, 27, 48, 51, 63, 67, 69, 70, 72, 74, 75, 77, 78, 86, 89, 90, 92, 96, 97, 98, 99, 103, 119, 145

H

human health, 3, 7, 48, 51, 100, 104, 105, 146

hydropower, 3, 85, 86, 87, 92, 93, 112, 113, 114, 115, 126

I

innovations, 16, 24, 25, 32, 36, 93

internalisation of costs, 52, 59

L

leader, 71, 74, 83, 119, 128, 129

local government, 13, 73, 76, 77, 80, 124, 132, 135, 136, 137, 145

M

management, 1, 12, 15, 17, 18, 21, 23, 27, 28, 34, 38, 39, 40, 42, 44, 61, 62, 73, 76, 80, 81, 82, 83, 90, 103, 108, 117, 118, 123, 129, 131, 132, 134, 135, 140, 141

N

natural environment, 1, 11, 70, 135, 141

Index

P

Poland, 1, 2, 3, 37, 38, 77, 88, 105–121, 123, 124, 125, 128, 129, 135, 137, 138, 142, 144, 145, 146

policy, 10, 15, 30, 33, 34, 35, 36, 41, 44, 47, 49, 58, 59, 63, 75, 76, 78, 79, 80, 86, 97, 100, 101, 103, 105, 119, 120, 136, 139

prosumption, 23, 128

public sector, 14, 40, 41, 46, 54, 62, 67, 99, 100, 118, 128, 130, 135, 136, 137, 139, 141

R

renewable energy, 1, 2, 3, 4, 6, 8, 12, 14, 15, 16, 17, 18, 22, 23, 37, 38, 44, 45, 47, 51, 59, 61, 62, 69, 70, 71, 73, 74, 75, 76, 85, 86, 88, 89, 90–104, 106, 107, 110–116, 118, 119, 120, 121, 124, 126, 127, 132, 133, 134, 135, 136, 137, 138, 139, 141, 142, 145, 146, 147

renewable energy sources, 1, 4, 14, 15, 17, 18, 22, 23, 38, 44, 45, 47, 59, 62, 69, 70, 73, 76, 86, 88, 90, 91, 92, 93, 94, 95, 96, 98, 99, 100, 101, 102, 103, 104, 106, 107, 111, 112, 113, 118, 119, 121, 124, 126, 134, 138, 139, 145

RES installations, 51, 73, 78, 79, 80, 92, 103, 112, 125, 127, 134

residents, 61, 73, 79, 108, 124, 129, 133, 134, 135, 136, 146

resources, 1, 3, 5, 6, 7, 9, 10, 13, 14, 27, 28, 29, 30, 33, 38, 39, 45, 48, 52, 53, 57, 59, 60, 61, 62, 63, 64, 65, 69, 80, 83, 86, 87, 90, 94, 95, 97, 99, 101, 103, 104, 105, 109, 110, 111, 112, 114, 116, 117, 123, 127, 128, 134, 135, 136, 138, 139, 140, 141, 143, 145, 146

S

security, 1, 4, 7, 12, 13, 18, 20, 24, 33, 34, 36, 44, 45, 46, 51, 59, 64, 73, 76, 83, 87, 90, 96, 100, 104, 119, 126, 136, 137, 141, 142, 143, 146, 147

society, 1, 3, 9, 10, 12, 13, 16, 24, 30, 32, 38, 42, 43, 49, 50, 51, 53, 55, 58, 83, 93, 95, 102, 119, 120, 134, 137, 139, 146

solar energy, 3, 85, 86, 93, 94, 97, 107, 112, 113, 114, 116

stakeholders, 3, 29, 47, 97, 98, 100, 102, 118, 125, 138, 139, 140, 141, 145

sustainable development, 3, 4, 7, 10, 12, 13, 36, 40, 69

T

thermal energy, 87, 93, 108, 110

transformation, 1, 11, 14, 34, 52, 74, 75, 77, 79, 93, 95, 102, 136

W

waste, 10, 29, 44, 46, 55, 61, 67, 85, 88, 92, 94, 97, 113, 116, 117, 118

wind energy, 85, 87, 92, 93, 95, 97, 107, 112, 113, 115